Stefan Schwab

Guaranteed Verification of Dynamic Systems

Karlsruher Beiträge zur
Regelungs- und Steuerungstechnik
Karlsruher Institut für Technologie

Band 12

Guaranteed Verification of Dynamic Systems

by
Stefan Schwab

KIT Scientific Publishing

Karlsruher Institut für Technologie
Institut für Regelungs- und Steuerungssysteme

Guaranteed Verification of Dynamic Systems

Zur Erlangung des akademischen Grades eines Doktor-Ingenieurs von der KIT-Fakultät für Elektrotechnik und Informationstechnik des Karlsruher Instituts für Technologie (KIT) genehmigte Dissertation

von Stefan Schwab geb. Maier, M.Sc.

Tag der mündlichen Prüfung: 22. Juli 2019
Hauptreferent: Prof. Dr.-Ing. Sören Hohmann
Korreferent: Prof. Dr. Vicenç Puig

Impressum

KIT Scientific Publishing

Karlsruher Institut für Technologie (KIT)
KIT Scientific Publishing
Straße am Forum 2
D-76131 Karlsruhe

KIT Scientific Publishing is a registered trademark of Karlsruhe Institute of Technology.
Reprint using the book cover is not allowed.

www.ksp.kit.edu

Print on Demand 2022 – Gedruckt auf FSC-zertifiziertem Papier

ISSN 2511-6312
ISBN 978-3-7315-0965-3
DOI 10.5445/KSP/1000097527

Preface

This thesis was written during my time at the Institute of Control Systems (IRS) at Karlsruhe Institute of Technology (KIT) and the department of Control in Information Technology (CIT) at FZI Research Center for Information Technology.

First of all, I want to thank Prof. Dr.-Ing Sören Hohmann for providing the necessary environment to allow successful scientific work like this. I really appreciate your never ending support and guidance throughout the years that enabled this thesis.
Also I'd like to thank Prof. Dr. Vicenç Puig for reviewing this thesis. Besides that, I really enjoyed our fruitful discussions that led to the collaborative development of the zonotopic method that forms a part of this thesis.

Further credits go to the IRS and FZI staff. I'll never forget this time and the vivid - not always work related - discussions. Also I'd like to thank all students and graduates that supported this work in some perspective.

Very special thanks go to Dr.-Ing. Gunter Diehm and Prof. Dr.-Ing. Mathias Kluwe for their very careful review of preliminary stages of this thesis. Your comments were very constructive, precise and rarely contradicting. You helped me very much to get the right focus and to add the correct final touch.

Last but not least I'd like to thank my wonderful wife Elisa. Without your support and understanding it would not have been possible to finish this work. Thank you for your constant optimism and for allowing me to spent that much time on science.

The very last sentences need to be understandable for two very special German boys:

Lieber Jonathan, lieber Samuel, das ist Papas Buch.
Es ist jetzt fertig.

Karlsruhe, in June 2019

There is no substitute for persistence!
It cannot be supplanted by any other quality.
With persistence will come success.

~ Napoleon Hill

Abstract

This thesis introduces a new specification and verification approach for dynamic systems. The introduced approach is able to provide type II error free results by definition, i.e. there are no hidden faults in the verification result. The approach is thus suitable to provide a reliable verification of safety critical systems.

A new notion of set based consistency for dynamic systems with a given specification is presented. Therefore Kaucher interval arithmetic is used to enclose the measurement data in a bounded error sense. The resulting method is able to verify the specified behavior of a dynamic system against its measurement data even in the presence of noise and sensor uncertainty. Consistency is defined using the Kaucher arithmetic united solution set which leads to mathematically guaranteed results.

It is proven mathematically that the desired property holds for a wide class of systems, including time invariant, interval type and hybrid systems, which can be used to describe even nonlinearities. Several extensions are introduced, leading to a new iterative identification and segmentation algorithm for hybrid systems which is able to handle even unknown switching times. In case the calculations can be done fast enough, the developed approach can also be used for the diagnosis of dynamic systems.

The presented methods are successfully applied to several example systems, including theoretic settings and a variation of different tank settings.

The new theories, methods and algorithms developed in this thesis form the foundation for reliable safety analysis of highly automated safety critical systems.

Zusammenfassung

Diese Arbeit beschreibt einen neuen Spezifikations- und Verifikationsansatz für dynamische Systeme. Der neue Ansatz ermöglicht dabei Ergebnisse, die per Definition frei von Fehlern 2. Art sind. Dies bedeutet, dass das Ergebnis der Verifikation keine versteckten Fehler enthalten kann. Somit können zuverlässige Ergebnisse für die Analyse von sicherheitskritischen Systemen generiert werden.

Dazu wird ein neues Verständnis von mengenbasierter Konsistenz dynamischer Systeme mit einer gegebenen Spezifikation eingeführt. Dieses basiert auf der Verwendung von Kaucher Intervall Arithmetik zur Einschließung von Messdaten. Konsistenz wird anhand der vereinigten Lösungsmenge der Kaucher Arithmetik definiert. Dies führt zu mathematisch garantierten Ergebnissen. Die resultierende Methode kann das spezifizierte Verhalten eines dynamischen Systems auch im Falle von Rauschen und Sensorungenauigkeiten anhand von Messdaten verifizieren.

Die mathematische Beweisbarkeit der Konsistenz wird für eine große Klasse von Systemen gezeigt. Diese beinhalten zeitinvariante, intervallartige und hybride Systeme, wobei letztere auch zur Beschreibung von Nichtlinearitäten verwendet werden können. Darüber hinaus werden zahlreiche Erweiterungen dargestellt. Diese führen bis hin zu einem neuartigen iterativen Identifikations- und Segmentierungsverfahren für hybride Systeme. Dieses ermöglicht die Verfikation hybrider Systeme auch ohne Wissen über Schaltzeitpunkte. Die entwickelten Verfahren können darüber hinaus zur Diagnose von dynamischen Systemen verwendet werden, falls eine ausreichend schnelle Berechnung der Ergebnisse möglich ist.

Die Verfahren werden erfolgreich auf eine beispielhafte Variation verschiedener Tanksysteme angewendet.

Dic ncuen Theorien, Methoden und Algorithmen dieser Arbeit bilden die Grundlage für eine zuverlässige Analyse von hochautomatisierten sicherheitskritischen Systemen.

Contents

List of Figures

List of Tables

Abbreviations and Symbols

Abbreviations

Abbreviation	Description
ARX	AutoRegressive System with eXogenous Input
CIT	Control in Information Technology
CMP	Center MisPlacement Effect
FS	Formal Specification
FZI	Research Center for Information Technology
HiL	Hardware-in-the-Loop
ILES	Interval Type Linear Equation System
IRS	Institute of Control Systems
KIT	Karlsruhe Institute of Technology
LMI	Linear Matrix Inequality Constraints
LTI	Linear Time Invariant System
PO	Prager-Oettli
PT1	Proportional Gain First Order Time Delay System
RRT	Rapidly Exploring Random Trees
SUT	System Under Test
VO	Verification Object

Symbols

Symbol	Description
$\square^*$	Nominal Value / Specification
$\square_{meas}$	Measured Values
$\square_{true}$	True Values of the VO
$\square_k$	Values at Time Step k
$\square_s$	Specific Sample from an Interval Value
$\square^{(i)}$	i-th Element of an Vector / Set
$\square^{-1}$	Inverse of a Matrix

Symbol	Description
$\square^{\dagger}$	Pseudo Inverse of a Matrix
$\square^{T}$	Transposed Vector or Matrix
$\hat{\square}$	Estimated Value
$\langle\square\rangle_{k=1}^{T}$	Time Series / Measurement Vector
α	Scaling Parameter
a_i	Input Parameter
a_n	Cross Section of Pipe n
A_n	Cross Section of Tank n
$\boldsymbol{A}$	Interval Type Regressor Matrix
$\boldsymbol{A}^{\mathfrak{C}}$	Interval Type Regressor Matrix with Arbitrary Assigned Quantor
$\boldsymbol{A}^{\exists}$	Interval Type Regressor Matrix with Assigned Exists Quantor
$\boldsymbol{A}^{\forall}$	Interval Type Regressor Matrix with Assigned Forall Quantor
$\boldsymbol{B}$	Interval Type Measurement Vector
$\boldsymbol{B}^{\mathfrak{C}}$	Interval Type Measurement Vector with Arbitrary Assigned Quantor
$\boldsymbol{B}^{\exists}$	Interval Type Measurement Vector with Assigned Exists Quantor
$\boldsymbol{B}^{\forall}$	Interval Type Measurement Vector with Assigned Forall Quantor
c_i	Output Parameter
$c_{\mathcal{M}}$	Constraints Given by the Measurement Data
$c_{\mathcal{N}}$	Constraints Given by the Nominal Set
$\mathcal{C}$	Consistent Set
δ	General Sensor Fault
δ_u^a	Maximum Absolute Sensor Fault on Signal u
δ_u^r	Maximum Relative Sensor Fault on Signal u
Δ_p	Passband Width
Δ_s	Stopband Width
Δt	Sampling Time
ϵ_k	Additive Noise of Measurement
$e_{n,k}$	Additive Noise of Model
$\boldsymbol{e}_a$	Neutral Element of Addition
$\boldsymbol{e}_m$	Neutral Element of Multiplication
$e^{(i)}$	Discrete Event
$\mathcal{E}$	Set of Discrete Events
f	Frequency
fl_{on}	Nominal Outflow of Tank n
f_f	Sensor Freeze Value
f_o	Sensor Offset Value
f_s	Scaling Fault
f_θ	Multiplicative Fault on System Parameter θ
$\mathcal{F}$	Feasible Set
$\mathcal{F}^{(k)}$	Feasible Set at Time Index k
$\mathcal{F}_k$	Feasible Set for all Measurement Data until Time Index k
$\mathcal{F}^{\square}$	Hyperrectangular Approximation of the Feasible Set
$\mathcal{F}^{\diamond}$	Zonotopic Approximation of the Feasible Set

Symbol	Description
$\mathcal{F}^{\circ}$	Polytopic Approximation of the Feasible Set
γ_n	Skaling Parameter of Tank n
g	Gravitational Force
$\mathcal{G}_Z$	Graph of the State Machine Z
h_n	Water Level in Tank n
h_{nml}	Height of Lower Valve, Connecting Tank n and Tank m
h_{nmu}	Height of Upper Valve, Connecting Tank n and Tank m
H^0	Zonotope Radius Matrix
$\mathcal{H}$	Hybrid System
$\mathbb{IR}$	Set of Proper Intervals
$\mathbb{IR}^*$	Set of Proper and Improper Intervals
$\Im$	Imaginary Part
k	Discrete Time Step
k_{det}	Detection Time of a Fault
k_{end}	End Time of a Segment
k_{err}	Activation Time of a Fault
k_{min}	Minimum Number of Necessary Measurement Points for the First Evaluation of an ARX System
k_{τ}	Switch
$k_{\tau'}$	End of Segment
k_p	Proportional Gain
$\boldsymbol{K}$	Unitary Interval Vector
$\mathbb{KR}$	Set of Improper Intervals
λ	Scaling Factor
$\boldsymbol{l}^{(i)}$	Activation Limit of Discrete Event $e^{(i)}$
μ	Mean Value
M	Fisher Matrix
n	System Order
n_a	Input System Order
n_c	Output System Order
n_e	Number of Events
n_q	Number of States
n_s	Number of Samples per Dimension
n_{check}	Number of Samples Used to Verify an Interval Type Specification
n_{switch}	Number of Switches
$\mathcal{N}$	Nominal Set
Ω_p	Passband Frequency
Ω_s	Stopband Frequency
p	Proportional Gain
$\boldsymbol{p}$	Interval Type Proportional Gain
P^0	Zonotope Center
q	Denominator Order
$q^{(i)}$	Discrete State

Symbol	Description
$\mathcal{Q}$	Set of Discrete States
Q	Penalty Matrix
$\boldsymbol{r}$	Arbitrary Interval Value
$\mathbb{R}$	Set of Real Numbers
$\Re$	Real Part
σ^2	Variance
s_k	General Measurement Signal
$s_{f,k}$	Signal with Freeze Fault
$s_{o,k}$	Signal with Offset Fault
$s_{s,k}$	Signal with Scaling Fault
$s^{(i)}$	Subsystem i
$\mathcal{S}$	Set of Subsystems
$\mathcal{S}_{con}$	Consistent Set of Subsystems
S	Penalty Matrix
$\sum$	Solution Set
$\sum_{\exists\exists}$	United Solution Set
$\sum_{\forall\exists}$	Tolerable Solution Set
$\sum_{\exists\forall}$	Controllable Solution Set
$\sum_{\forall\forall}$	Strong Solution Set
$\sum_{a}$	Algebraic Solution Set
θ	System Parameter
θ_{com}	Common System Parameter
$\theta_{err,k}$	Faulty System Parameter
$\boldsymbol{\theta}$	Interval Type System Parameter
Θ	System Parameter Vector
T	Measurement Time
T_{calc}	Calculation Time (Runtime) of the Algorithm
$\tilde{T}$	Delay Time
u	Input
u_{mean}	Mean of Signal u
U^*_{init}	Initial Input Values
φ	Measurement in Regressor Form
v_n	Measurement-Signal of Pump n
v_{outn}	Nominal Outflow Valve of Tank n
v_{nml}	Lower Connection Valve between Tank n and Tank m
v_{nmu}	Upper Connection Valve between Tank n and Tank m
v	Vertex
v_{dec}	Index of a Vertex in Decimal Value
V_{bin}	Index of a Vertex in Binary Value (Vectorial)
$\mathcal{V}$	General Set of Vertexes
$\mathcal{V}^{\square}$	Vertexes Defining a Hyperrectangle
$\mathcal{V}^{\diamond}$	Vertexes Defining a Zonotope
$\mathcal{V}^{\pentagon}$	Vertexes Defining a Polytope

Symbol	Description
W	Enabler Signal
$\underline{x}$	Infimum of Interval Variable
$\overline{x}$	Supremum of Interval Variable
x_c	Center of Interval Variable
x_Δ	Radius of Interval Variable
$x^\ominus$	Negative Part
$x^\oplus$	Positive Part
x^+	Magnitude
$\boldsymbol{x}$	Interval Variable
$\boldsymbol{X}$	Interval Type Solution Vector
y	Output
Y_{init}^*	Initial Output Values
Z	State Machine
$\mathcal{Z}$	General Zonotopic Set

1 Introduction

The fast technological development in computer engineering in recent years led to very powerful computing capacities that are now available at very low costs [Wil17]. As a result those chips are used in an increasing number of products to make them "smart" and to enhance user experience and functionality. These smart devices are interleaving the daily life of millions of people and are used for an increasing number of tasks [Gho17]. State of the art techniques are powerful and mature enough to take over even very complex and sensitive tasks - for example in autonomous cars, in flight assistance systems or in the control of critical infrastructure. Tasks that can potentially harm human beings or destroy valuable infrastructure are called "safety critical" and special measures need to be taken during the development cycle to ensure correct operation of such safety critical systems [IEC10][ISO11].

These special measures are given by safety analysis methods. A very relevant property of all safety analysis methods is given by the amount of their type I and type II errors. Thereby type I errors (false alarms) denote the situation in which a safety analysis method evaluates a correct system to be faulty. The complementary condition is given by type II errors (hidden faults). In this case a safety analysis method evaluates a faulty system to be correct. Type II errors are of major importance in the context of safety critical systems. A faulty system that is evaluated to work correctly poses uncontrollable risk to the user and the environment. Thus there is a need for safety analysis methods that do not suffer from type II errors.
In the context of this thesis, verification of dynamic systems means applying safety analysis methods in an offline setting to ensure consistency of the verification object (VO) with the specification. Guaranteed verification means that type II errors are impossible by design. In case the safety analysis is fast enough, it can be applied in an online setting which is then called "diagnosis". Diagnosis can also be used to detect runtime errors.
It is common opinion that there is currently no sufficient type II error free method available in the state of the art and the state of science [Kap16].

Currently safety analysis methods use mostly falsification approaches, e.g. methods from the field of testing. To achieve confidence about the absence of failures based on testing methods it is necessary to use a sufficiently large amount of test cases. This leads to the fact that safety analysis is more expensive as the development itself [Fos15] and costs are expected to rise further with increasing complexity of the tasks assigned to the technical system.

Besides costs, current safety measures are often based on the experience of the responsible engineer or brute-force simulation approaches are applied [ZN09]. It is widely recognized that those methods will not keep up with the complexity given by currently developed or future systems [Ram17][Ott18].
The example which is predominant in public perception is located within the automotive industry. Current systems like automatic cruise control or autonomous parking pilots are analyzed by applying the previously mentioned experience and simulation based safety methods [Zan12]. Nevertheless it is known that those are not suitable for the arising challenges, e.g. in the context of autonomous driving [Wac17][Koe18].

A possibility to avoid this dilemma is to use formal methods that can proof specific properties of a verification object. A promising approach that gained great attention in recent years is given by interval arithmetic safety analysis methods, see among others [Uga03][Wol10] [San17]. Results obtained using those method are guaranteed to include all possible nominal system behavior as well as additional non-nominal behavior (so called spurious solutions). Due to this overapproximating property, there are no type I errors. However, for the same reason type II errors are possible by design.

The goal of this thesis is to close the gap by developing a formal method for the safety analysis of dynamic systems that is guaranteed to be free of type II errors.

2 State of Science

There are numerous methods and approaches concerned with (safety) analysis of development results in different communities. Also, there is a broad terminology with respect to the verification question. The primary goal of this chapter is to build a basic conceptualization and the resulting terminology used in this thesis. This is necessary to follow the ideas and approaches introduced in later chapters.
Furthermore, the most important and wide spread notions and methods used in engineering and engineering science are introduced and discussed.

2.1 Conceptualization and Terminology

System behavior analysis can be conducted with respect to different perspectives. This thesis addresses the verification of dynamic systems. A classification of other perspectives is given in Appendix A. To conduct the verification of dynamic systems it is necessary to define three components. First a description of the intended system behavior is set up. The next step is to define the concept of deviating behavior. Finally the developed real system behavior has to be assessed with respect to the intended behavior.

2.1.1 Behavior Description

The desire of the costumer needs to be documented in some kind of specification to allow any analysis in terms of verification. There are as many specification methods to define the nominal behavior as there are methods to check their fulfillment. The variety includes very formless approaches in human language [Mac95] as well as very formal definitions using (runnable) models [AI15] or special specification languages ([Par72][Spi89][Abr96]). The choice of a suitable specification formalism to be used in a project is a trade-off. The less formal a specification, the less effort is necessary to set it up, leaving the effort to the developer who needs to interpret the specification. During the verification procedure it is necessary to interpret the specification which leads to a need for experienced experts [ZN09, p. 120][Raj13][Bal16]. The more formal a specification, the more effort is necessary to set it up. An advantage of formal specifications is that they force the specification engineer to capture the requirements in a precise and structured way. Standardized specification routines help to avoid careless mistakes during the setup [Par86][Hal90][Sch15a].
On the other hand all properties that should be covered need to be representable in the specification, which can lead to requirements being impossible to be captured in a specific formalism. However, due to the precision of very formal specifications it is possible to analyze them in a rather automated or "proof-like" way.

Throughout this thesis it is assumed that the specification itself is known, correct and represents the whole functionality, behavior and all properties that are necessary to fulfill the customers desire.

This work assumes a set based specification. It is assumed that it is possible to represent the desired behavior of a dynamic system in terms of a specific abstract set. This formal specification (FS) can be interpreted to include all dynamic system parametrizations that are able to create the intended behavior. The set of intended or desired behavior can be given by differently shaped sets e.g. a circle or a square. In case there is no variation in the desired behavior, the set consists only of one parametrization which is given by a distinguished point. Different possible specification sets are depicted in Fig. 2.1.

Figure 2.1: Exemplary set based specifications (Point, Circle, Square)

2.1.2 Behavior Deviation

Implemented systems can show behavior deviating from the desired behavior due to several causes. From a very basic point of view it is possible to differentiate between mistake by misfortune, mistake by accident without intention and mistake by deliberate wrong-doing by an individual.[1]
The setting of this work tackles the second kind, mistake by accident without intention that can happen at every point during the development process. A wide range of expressions is used to differentiate the field of unintended behavior or unintended properties. However, different fields of research and profession are using different naming conventions.
The naming convention used in this thesis is given in Fig. 2.2.

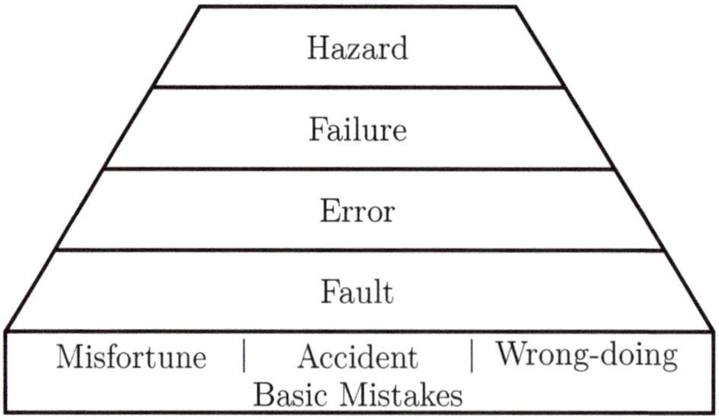

Figure 2.2: Failure terminology

[1] This categories are inspired by Aristotle (384 - 322 BC) who thought about ethics and mistakes of human behavior [Res07].

The foundation of all unintended behavior is given by the basic mistakes. The next instance is called fault and denotes the deviation of at least one system value from its intended value.This deviation can happen due to all three of the basic mistakes. If a fault leads to unintended system behavior it is called error.
A system perturbed by a fault and a resulting error can still be able to operate correctly. Only if there is a persistent interruption of correct behavior the system is called to show a failure. This failure introduces a hazard into the environment the system operates in. The hazard can lead to consequences in the environment that potentially harm objects or even human beeings.

Complementary, there is the concept of disturbance in a control engineering sense. The process of capturing real world data and transferring them into any control system is always superimposed by a process that creates a deviation between the real values and the measurement values [Fra16, p. 65]. This deviation is called disturbance or noise and every system needs to be adapted to the specific noise present in itself as well as in the particular environment.

2.1.3 Behavior Assessment

The assessment of the verification object is done with respect to its behavior. Therefore it is necessary to set up the formal specification (FS) and additionally describe the behavior of the verification object (VO) using the same formalism. Both descriptions are assumed to be represented by a convex set. The notion of set based basic consistency that is used throughout this thesis is given in Definition 2.1.

Definition 2.1 (Set Based Basic Consistency)
A set based verification object VO is called basic consistent with its set based formal specification FS if and only if there is an intersection between the formal specification and the verification object behavior:

$$(FS \cap VO \neq \emptyset) \Leftrightarrow Basic\,Consistency.$$

A special case is given by full consistency, which means that all behavior given by the formal specification is available in the verification object.

Definition 2.2 (Set Based Full Consistency)
A set based verification object VO is called full consistent with its set based formal specification FS if and only if the formal specification behavior is an subset of the verification object behavior:

$$(FS \subseteq VO) \Leftrightarrow Full\,Consistency.$$

The inverse is given by inconsistency according to Definition 2.3.

Definition 2.3 (Set Based Inconsistency)
A set based verification object VO is called inconsistent with its set based formal specification FS if and only if there is no intersection between the verification object behavior and the formal specification behavior:

$$(FS \cap VO = \emptyset) \Leftrightarrow Inconsistency.$$

This means that none of the available VO behavior is given in the specification.

The resulting situations are depicted in Fig. 2.3. It is assumed that the formal specification FS is given by the blue circle. The set of real VO behavior is given by the green and red circles. Definition 2.1 and Definition 2.2 are fulfilled in the left and middle subfigures, leading to the verdict $Basic\ Consistency$ and $Full\ Consistency$ for the depicted VO and the FS. Definition 2.3 is fulfilled in the right subfigure, leading to the verdict $Inconsistency$ for the depicted VO and the FS.

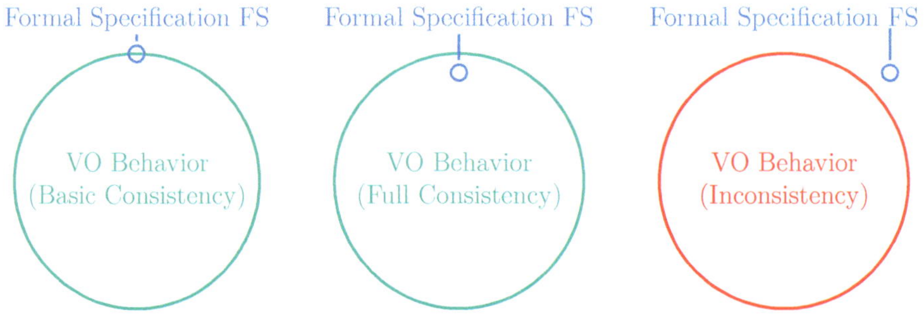

Figure 2.3: Basic consistent, full consistent and inconsistent result of set based verification

In the context of this thesis the verification object is considered to be correct if there is specified behavior within the VO behavior. This is called "consistent behavior".
For the ease of notation, the term $Consistency$ is used as soon as there is "consistent behavior", either due to $Basic\ Consistency$ or due to the even stricter $Full\ Consistency$. The following considerations apply equivalently to both definitions.

In general, the VO behavior is not directly available and thus needs to be captured by an approximation. If the behavior is given in terms of dynamic system parameters, the approximation can be calculated using identification methods [Lju99]. Therefore it is necessary to interact with the real VO to determine the underlying behavior. Assumption 2.1 has to hold to allow a successful identification.

Assumption 2.1 (Persistent Excitation of the VO)
The verification object VO is sufficiently excited to show all relevant behavior.

Only behavior of the VO that is triggered or excited is included in the approximation and can thus be analyzed [Ast95, p. 41][Ise10, p. 250]. Throughout this thesis it is assumed that Assumption 2.1 holds.

There are two main set based calculation paradigms that can be used to determine the approximation of the VO: underapproximation ($^-$) and overapproximation ($^+$).
In case of overapproximation, there is spurious behavior in the resulting outer enclosure (see rectangle in the left of Fig. 2.4). If underapproximation is used, some VO behavior is missing in the inner enclosure (see rectangle in the right of Fig. 2.4).

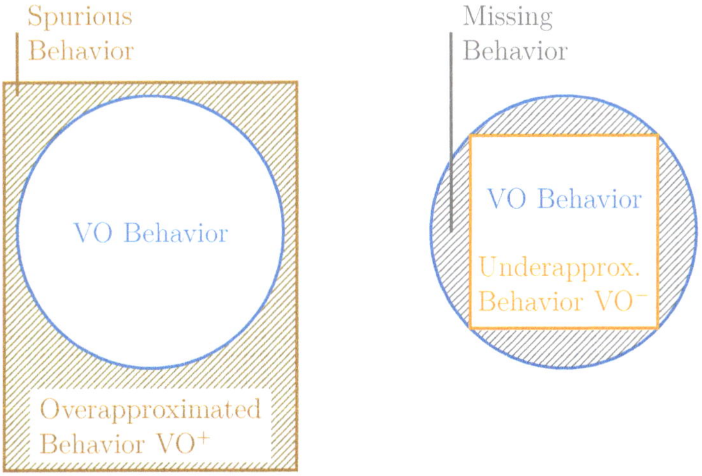

Figure 2.4: Set based overapproximation and set based underapproximation

As the true VO behavior is not available, the approximated behavior is used to reason about $Consistency$. In case an overapproximation of the verification object behavior (VO$^+$) is used, this leads to

$$(FS \cap VO^+ \neq \emptyset) \Leftrightarrow Consistency^+. \tag{2.1}$$

It is very important to note that (2.1) yields the verdict $Consistency^+$ that holds only for the overapproximated behavior VO$^+$. It does not have to be valid for the true VO behavior. To relate the verdict with the true VO behavior, the overapproximating property

$$(VO \subseteq VO^+) \Rightarrow Consistency \subseteq Consistency^+ \tag{2.2}$$

has to be taken into account, leading to

$$(FS \cap VO^+ \neq \emptyset) \Leftrightarrow Consistency^+ \Leftarrow Consistency \Leftrightarrow (FS \cap VO \neq \emptyset). \tag{2.3}$$

It can be seen from (2.3) that the $Consistency^+$ verdict can not be extended to the true system behavior VO. This situation is depicted in the lower left field of Tab. 2.1. This property holds for both, $Full\ Consistency$ and $Basic\ Consistency$.

If an overapproximating method yields the result $Consistency^{+}$ and this result is generalized to $Consistency$, it is possible that there is a "hidden fault" present in the system. This situation is called type II error and it is a severe problem in the field of safety critical systems. Type II errors are likely to harm people or the environment as the VO is showing wrong behavior but the supervising system assumes correct functionality. Corrective actions that are designed to prevent fault induced damage are not activated in case of a hidden fault. This type of fault can thus proceed and potentially harm people. In safety critical systems, type II errors need to be avoided under all circumstances.

Therefore it is beneficial to use the underapproximation of the verification object behavior (VO^{-}), leading to

$$\left(VO^{-} \subseteq VO\right) \Rightarrow Consistency^{-} \subseteq Consistency. \tag{2.4}$$

The resulting verdict $Consistency^{-}$ can be extended to the true system property:

$$\left(FS \cap VO^{-} \neq \emptyset\right) \Leftrightarrow Consistency^{-} \Rightarrow Consistency \Leftrightarrow \left(FS \cap VO \neq \emptyset\right). \tag{2.5}$$

If the underapproximation of the verification object VO^{-} is consistent with the specification, it is guaranteed that the true verification object is also consistent with the specification (see Tab. 2.1, lower right field). Thus the verdict is guaranteed to be free of type II errors. Again this property holds for $Full\ Consistency$ and $Basic\ Consistency$.

The new verification approach developed in this thesis is based on the concept of underapproximation to avoid type II errors by design. This essential property is necessary to solve the currently unsolved reliable verification problem of safety critical systems.
However, the property comes at the costs of possible false alarms (see Tab. 2.1, upper right field). Instead of additional spurious solutions there are missing solutions generated by the underapproximation VO^{-}. Even though there is consistent behavior in the VO, this behavior is not included in the underapproximation, leading to a false alarm (type I error).
The situations in the remaining upper left field of Tab. 2.1 depicts the correct verdict $Inconsistency$ than can be obtained using both, under- or overapproximation. This is due to the fact that there is no consistent behavior for the real VO as well as for both approximations.

Table 2.1: Error types

	System is Faulty	System is Correct
Verification Result: “Inconsistency”	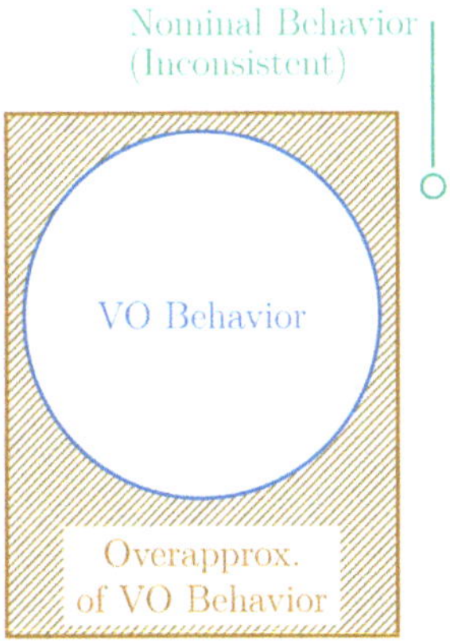 Correct Result	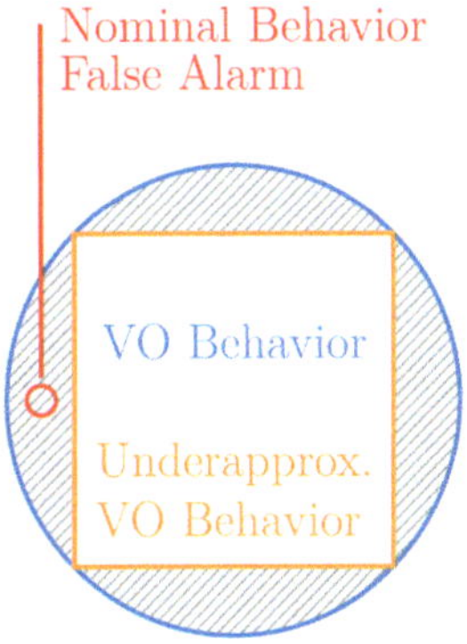 Type I Error
Verification Result: “Consistency”	 Type II Error	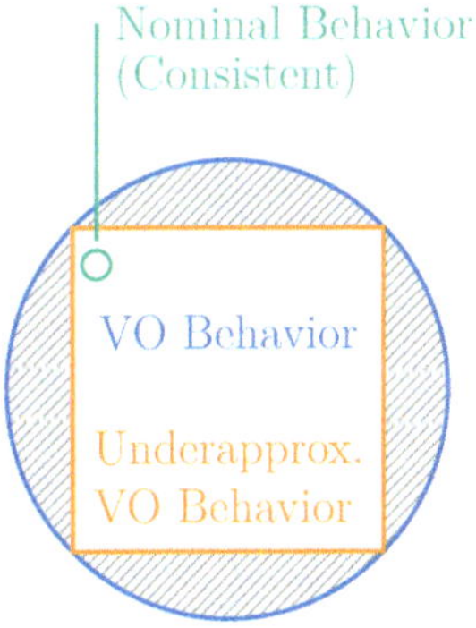 Correct Result

2.2 Interval Arithmetic Methods

The overapproximating property introduced in the previous section can be achieved using the notion of interval arithmetic. Interval arithmetic is thereby used to enclose the effects of noise and epistemic lack of knowledge that is always present in real systems. A fixed lower and upper bound is used to describe a set of possible true values x_{true} that are associated with a given measurement value x_{meas}:

$$x_{true} \in [x_{meas} - \delta,\ x_{meas} + \delta]. \tag{2.6}$$

The maximum tolerance δ is the only parameter needed to set up the interval. This interval arithmetic notation of maximum deviation is widely used by sensor manufacturers [Kre95] and in the fault detection community (e.g. in [Arm09][Zai14][AI17]). The true measurement value is guaranteed to be included in the interval and it is guaranteed that the true value is never outside the interval.
When interval enclosure is used on the measurement data, all succeeding calculations have to apply the notions of interval arithmetic to preserve the guaranteed properties. The basic property of interval arithmetic calculations is that all possible solutions are included in the result. Therefore, interval arithmetic results are able to create the introduced overapproximation properties. The interval arithmetic solution set consists of real solutions and spurious solutions. Spurious solutions denote solutions that do not exist in the real system but are inevitable artifacts that are introduced by interval arithmetic calculations and the final interval arithmetic (and thus axis parallel) enclosure of the real solution [Bau87].

Interval arithmetic is widely used for verification [Bal16] and diagnosis methods as type I errors (false alarm) are prevented by definition. Therefore, models of the nominal system are used to calculate a set of predicted outputs for the measured inputs of the system [Ven15]. This can be done by using intervals on the system parameters to calculate an interval range of outputs [Pui06][Mes10][Wol10]. The system is assumed to be correct as long as the measured output values are within the predicted output interval, i.e. within the so called direct image. Due to the used outer enclosure of the prediction, type II errors are possible using this class of methods.
An alternative approach uses the so called inverse image [Pui06] or feasible set [Cas14], also known as set-membership approach [Ing09]. In this case the input-output measurement data is used to calculate the set of parameters that is able to generate the observed mapping. This is possible if the system is linear or nonlinear but linear with respect to the parameters. If there is a member of the set of nominal models within the feasible set of the measurement, the measured data can be explained by the nominal model. This class of methods utilizes interval arithmetic identification based on outer enclosures. Therefore there are type II errors possible by definition.
The feasible set resembles the solution set of the identification problem given by the measurement data that can be computationally hard to calculate [Hor13].
A wide spread possibility to approximate the solution set is given by subpavings using the SIVIA algorithm (see [Jau01, p. 45ff][Pui06][Mes10]). The bisection approach of SIVIA leads to a large set of different intervals with various size. Even though this result is very precise, the great amount of intervals leads to a complex handling and to long calculation times.

A more efficient approach is given using a zonotope[2] representation as shown in [Ing09]. The question how this approximation can be calculated in an efficient way is still an active research topic. Latest results [Koc19] use sparse polynomial zonotopes. Additionally, there are methods to reduce the solution set by pruning spurious parts if possible. For example the approach presented in [Wol10] uses measurement data to reduce the overapproximated state set. Nevertheless, all approximation methods use outer enclosures and are thus prone to type II errors.
Besides the field of diagnosis, different interval arithmetic approaches are used for state estimation ([Jau01][Ram09][Mes10][Efi13][Kre16][Kre18][Wan18]) and control [Rau06]. None of these methods addresses type II errors.
Therefore it is necessary to develop a new verification method that is able to guarantee the absence of type II errors. This can be achieved by calculating an inner enclosure that consists of a subset of the real solution as shown in the previous section. This thesis utilizes a special extension of interval arithmetic called Kaucher interval arithmetic that provides powerful theories to calculate the necessary inner enclosures.

2.3 Governing Complexity: Time Variant and Hybrid Verification Approaches

Linear time invariant (LTI) systems form the base for the most system theoretic methods and approaches. Nevertheless, real world systems are normally neither linear nor time invariant. It is thus the question how to handle the complexity inherently given in real world systems. Some nonlinear systems can be handled using nonlinear theory which provides methods for analysis and control [Kha15]. However those approaches are often subject to very strict preliminaries and only applicable for a narrow class of systems. Therefore linearization is often used in practice to handle nonlinearities. Another possibility is to model nonlinearities by using time variant parameters in a linear system [Ble11]. In this case, the set of feasible parameters can be bounded using interval arithmetic or any other set definition. It is also possible to split the nonlinear dynamic into a sequence of linear dynamics [Oza14] that are activated by a superimposed switching mechanism. The resulting model belongs to the class of hybrid systems. Hybrid systems can also be used to model time variant linear systems with piecewise constant parameters. Therefore the different piecewise constant parameters are represented using an individual dynamic subsystem each.

There is a wide theoretical framework available for hybrid systems [Eng02][Mah10]. The topic of hybrid verification is subject to current research in the cyber physical systems community (see e.g. [Sch15b][Kap16][Sch17a][Ara17] [Bar18][Har18][Lau18]). There are also large research clusters in this area [AVA19][ENA19].
To apply the set based approach introduced in this thesis in a hybrid setting, it is necessary to use hybrid identification methods to determine the unknown system parameters from given measurement data. The most relevant hybrid identification approaches given in the literature are introduced in the following.

2 A zonotope is a convex polytope that is point symmetric with respect to its center.

The algebraic approach provided in [Vid08] interprets the identification setting as a geometric problem. The measurement data as well as the parameters are interpreted as vectors that have to be perpendicular in case the identified parameters match the true parameters. The goal is to find the parameter vector with minimal projection on the family of all measurement vectors. A Bayesian approach based on stochastic properties is given by [Jul05]. The number of models and the model order need to be known a priori for this procedure. A cluster based approach was developed by [FT03] were machine learning methods are used to form groups of similar behavior. Identification methods based on optimization were developed and presented in [Mün05][Bor09][Lau18]. The bounded error approach introduced in [Bem05] and used for time variant systems in [Bra16] assumes errors that are characterized by their maximum value. Even though this is close to the basic interpretation used in this thesis, those approaches do not use interval arithmetic notations. Therefore they lack the guarantees that are necessary in the safety critical context of this thesis.
A greedy approach based on [Oza12] was developed in [Die13a] and [Die13b]. This approach is different from the others as it is the only one that uses a multi-step prediction error instead of the common one step prediction error. It was extended to cyber-physical systems in [Sch17a] and to Kaucher interval arithmetic in [Sch19].

2.4 Other Common Verification and Falsification Approaches

There is a wide range of verification methods with different degree of abstraction and formalization.
A strong diagnosis community is active in the control engineering field (see e.g. [Ise93] [Sch03][Ven03a][Ven03b][Ven03c][Bla06][Arm09][Sch09][Pui10][Ise11][Cac13][Zol14]).
Also, a large testing and verification community formed in the information technology community. This leads to a great range of verification and validation methods, unfortunately using a similar terminology (e.g. [Bar78][Boe84][Hay86] [TF91][Bar05]). Three of the most relevant approaches are introduced in this section.

2.4.1 Testing

A wide spread - if not the most wide spread - approach to assess the properties of a technical system is given by testing. Testing is a classic falsification method, aiming on the detection of counter examples that do not show the intended behavior. It is therefore necessary to define a well-chosen set of test cases, consisting of the system state at the beginning of the test case, inputs that are applied during the run of the test case and outputs that are expected to appear during or at the end of the test case. If the system under test (SUT) shows outputs deviating from the expected outputs, the test case is called a "failed test" and further inspections of the test case are necessary to identify the reason. In contrast to verification methods, each result is only valid for the specific applied test case. It is long known that the confidence of the results can only be increased by increasing the amount of test cases [Fut89, p. 3].

Testing can be applied at different stages of the development cycle and at different abstraction levels. The state of the art testing scheme is given by the V-model (see e.g. [Web09][ZN09][Raj13][Ott18]) as depicted in Fig. 2.5.

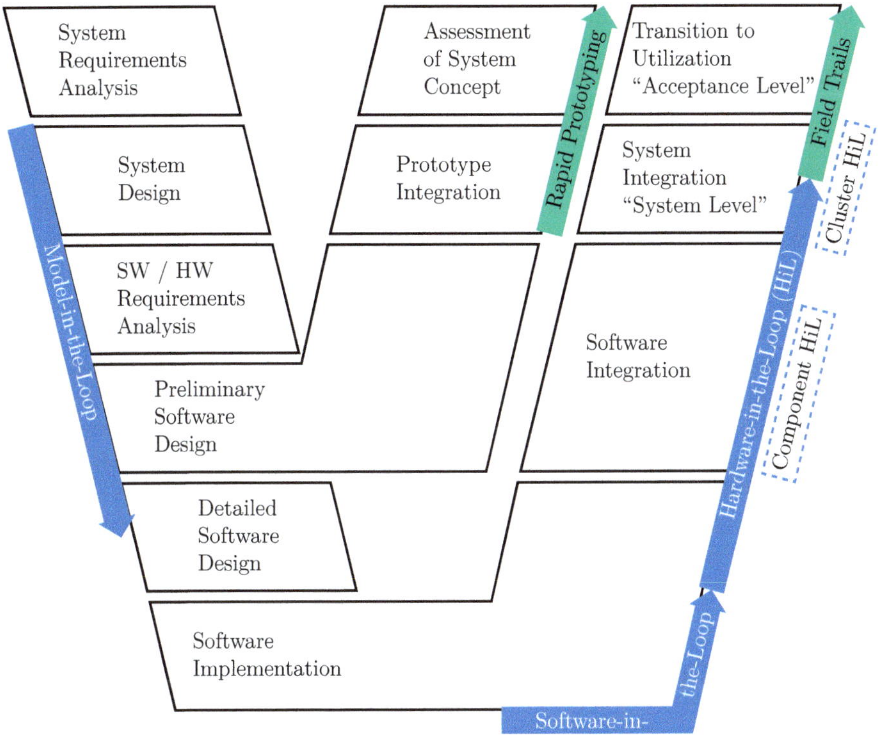

Figure 2.5: V-model diagram (based on [Ott18])

The left part of the V-model is the specification branch. It is applied in a top-down fashion. The specification is initialized on the highest level representing the customers desire. Then it is propagated to lower levels and refined to match the degree of formalization of each level. This structured procedure includes the decomposition of the overall task in several sub tasks accompanied by the definition of interfaces between the sub tasks.

After every refinement step the result is checked against the superimposed level to verify that both requirements are consistent. The final (software) specification is implemented at the bottom level. The resulting SUT is checked against its specification at the same level. If the test succeeds, the function is used in higher levels and combined with other parts to meet superimposed specifications. If a test fails, the system under test is rejected to the previous level. The complexity of the SUT increases with rising level on the right verification branch of the V-model. It is also possible that a failed test at high levels (e.g. at system or acceptance level) leads to changes in the respective high level specification. This results in the need of repeating the whole development process starting with the changed high level specification [Raj13][ZN09].

There are several technologies that are used in different phases of the V-model, sometimes leading to additional branches. Fig. 2.5 includes cutting edge technologies like rapid prototyping, component HiL and cluster HiL that are used to speed up the time effort to perform a complete cycle of the V-model.
An important part of the testing process is the test case generation. There are several possibilities to set up the test cases. If there is experience with this kind of product, it is likely that an existing test case database can be reused and adapted [Sax08][Bal16]. Another straight forward approach is to determine the feasible range of all input variables and divide this range in so called "equivalence classes" [Utt06]. The test cases are then formed by combining one representative (or the minimum and maximum values) from each equivalence class with all possible representatives of the other input variables [Utt06]. If it is not possible to form equivalence classes, it is also possible to sample the valid range of a variable randomly or in equidistant steps. This approach is straight forward and easy to understand but it will lead to very large sets of test cases with increasing number of variables in the system or if a fine resolution of the variable ranges is necessary (i.e. [Hei05][AI15]).
The runtime for testing or simulation rises with the number of test cases, leading to large computation times. If equidistant sampling is used, it is possible that much calculation time is spent assessing "uninteresting" regions of the input space or that "interesting" regions are only covered with few test cases. One major drawback of the testing approach is that all test cases need to be redone if the SUT changes. As there are frequent iterative changes during a development cycle (i.e. more than one iteration of the V-model is necessary), the test cases need to be applied several times which leads to even longer computation times.

2.4.2 Reachability Analysis

By including more system theoretic knowledge, testing can be developed to reachability analysis. The basic requirement for this purpose is a specification that includes some kind of state space the system is operating in. Further, the specification has to define forbidden areas in this state space.
The main idea of reachability analysis is to determine whether there is a sequence of inputs that leads the SUT to enter the forbidden area. If an input sequence leading to a forbidden area is found, the SUT is falsified and needs to be improved to match the requirements (see among others [Bha04][Alu06][Mit07][Alt08][Don10]). Reachability analysis is also aiming on finding a counter example which has the same basic problem as the testing scenario: no detected counter example does not mean there is no fault present in the system as faults can be hidden in uncovered parts of the state space.
Runtime limitations restrict all methods to a finite number of samplings and thus to partial coverage of the state space. There are smart coverage criteria available that allow an effective calculation of the most important regions of the state space e.g. using rapidly exploring random trees (RRT) (see i.e. [Bha04][Kap16][Pan17]). These approaches can be extended to the so called method of star discrepancy [Dan11] or by applying the underminer method [Bal16]. Nevertheless, reachability analysis is still a falsification method, based on the specification of the faulty case (forbidden areas). Therefore reachability analysis is not suitable to solve the verification problem addressed in this thesis.

2.4.3 Formal Verification

One possibility to avoid the counter example problem is given by more abstract approaches from the verification field. Those formal methods are used to reason about system properties in a mathematical rigorous way. To apply formal methods, the VO needs to be transferred to a strict mathematical notation, e.g. by using the *Z* specification language (see e.g. [Spi89][Bro05, p. 325]) or *Prolog* [Bro05, p. 334]. The formalized VO can then be used to carry out mathematical proofs showing that specific system properties hold in all operating conditions. The proof is thereby conducted by a so called theorem prover.
This approach is very powerful as the results are valuable and mathematically sound. Nevertheless, formal proofs can only be done for very distinct properties. Furthermore the methods need very long runtime even for "small" academic problems. This leads to still unsolved runtime issues for real world problems ([Bro05, p. 325][Bar18]).
A basic problem that cannot be omitted is that formal proofs cannot be conducted on the VO directly. Therefore the results hold only for the image of the VO which is given in the used formalism. Mistakes that are introduced when the system is transferred from the real world into the modeling formalism cannot be detected.

2.5 Scientific Gap and Related Research Question

Testing is the state of the art for current systems and is successfully applied in various communities. Nevertheless there are current systems, e.g. autonomous driving functions, that show a number of relevant scenarios that cannot be covered by testing or simulations. Even if this was possible, falsification methods cannot prove the absence of all faults. They need to stop at some point and have to assume that no undiscovered fault is present in the system. There are established methods available that use interval enclosures to mathematically bound the system behavior. Those methods use classic interval arithmetic, leading to overapproximating properties. Overapproximating methods are able to provide type I error free results, meaning that there are no false alarms generated by the method. Nevertheless, the overapproximation can cover missing behavior, leading to an undetected hidden fault. In the case of safety critical systems, hidden faults (type II errors) can lead to severe consequences threatening human life. It can be concluded that the verification of safety critical dynamic systems is currently not solved.

A new verification method has to be developed to close this gap. This method has to be free of hidden failures, meaning that there are no type II errors. Therefore the specification is assumed to be formally given in terms of a set of dynamic system parameters. The behavior of the VO has to be given in the same formalism, leading to an identification problem. Safety critical systems are often implemented as embedded systems that consist of a closely connected discrete event system (the controller) and a dynamic system (the plant). Thus it is necessary to develop a hybrid identification method that is able to provide the desired guarantees.
The comprehensive research question tackled by this thesis is:

"How can the consistency of highly automated safety critical dynamic systems be evaluated by a guaranteed verification method?"

3 Methodical Approach and Mathematical Preliminaries

Considering the state of science as well as the current and future challenges of system theory there is a need for a new verification method. The rising importance of safety critical systems emphasizes the need for formal methods that target type II errors. This thesis introduces such a formal method based on the notions of interval arithmetic, extended to Kaucher interval arithmetic. First the necessary notations and definitions are given to provide a sound theoretical base for further considerations.

3.1 Mathematical Preliminaries

All methods introduced in this thesis are based on interval arithmetic, appended by the properties of Kaucher interval arithmetic. In the following, the basic properties and notations of interval arithmetic are introduced.
The goal of this chapter is to provide a brief overview of interval arithmetic that is necessary for this thesis. The interested reader is referred to [Bau87][Rze08][Roh12][Sai14] for an extensive coverage of the topic. Throughout this thesis the well known notation of interval arithmetic extended by Kaucher interval arithmetic introduced in [Kup95][Sha96] is used.

3.1.1 Basic Interval Arithmetic

Interval arithmetic was initially developed to handle numerical calculation errors due to floating point calculation used in computer algebra systems [Apo67]. It gained popularity outside the numerical community with the rise of electronic computing in various fields. When measurement data is used in computing - as it is normally the case in engineering and natural science - faults are already created by the measurement process itself [Kre95]. Furthermore, the used values are given as samples at discrete time steps k. Every measurement $y_{meas,k}$ is compromised by some noise ϵ_k that leads to a deviation between the real value $y_{true,k}$ and its measurement

$$y_{meas,k} = y_{true,k} + \epsilon_k. \tag{3.1}$$

It is possible to define intervals around the measurement that are guaranteed to include the real system value

$$y_{true,k} \in [y_{meas,k} - \delta,\ y_{meas,k} + \delta] \tag{3.2}$$

if the maximum δ of the absolute noise is known i.e. $\forall k : |\epsilon_k| \leq \delta$.

Suitable values of δ can be determined from the data sheets provided by the sensor manufacturers.

The definition of a classical interval type variable $\boldsymbol{x}$ as given in [Sai14] is

$$\boldsymbol{x} := [\underline{x},\, \overline{x}] = \{x \in \mathbb{R} \mid \underline{x} \leq x \leq \overline{x}\}\,. \tag{3.3}$$

This definition includes all real numbers that are between or on the infimum $\underline{x}$ and the supremum $\overline{x}$. In case $\underline{x} > 0$ and $\overline{x} > 0$ the interval is called positive interval. If the infimum is negative ($\underline{x} < 0$) and the supremum is positiv ($\overline{x} > 0$) i.e. if $0 \in \mathbf{x}$, the interval is called zero interval. A negativ interval is given if $\underline{x} < 0$ and $\overline{x} < 0$. One last definition covers the case of supremum and infimum being the same, i.e. $\underline{x} = \overline{x}$, which is called a degenerated interval [Dja17] or point real interval [Sai14].
The set of so called proper intervals is given by

$$\mathbb{IR} := \{\boldsymbol{x} = [\underline{x},\, \overline{x}] \mid \underline{x} \leq \overline{x} \text{ and } \underline{x}, \overline{x} \in \mathbb{R}\}\,. \tag{3.4}$$

Despite this infimum-supremum notation, each proper interval can be given using the center

$$x_c := \frac{1}{2}\left(\overline{x} + \underline{x}\right) \tag{3.5}$$

and the radius

$$x_\Delta := \frac{1}{2}\left(\overline{x} - \underline{x}\right). \tag{3.6}$$

of an interval. The interval can now also be stated in the center-radius notation

$$\boldsymbol{x} = \langle x_c,\, x_\Delta \rangle\,. \tag{3.7}$$

Furthermore, it is important to introduce the interval type vector matrix notation based on [Jau01]. Vectors and matrices are written as capital letters X and interval values are given in bold font $\boldsymbol{x}$, leading to interval matrices denoted as $\boldsymbol{X}$. An interval vector $\boldsymbol{X}$ is defined as cartesian product of n closed intervals that includes a subset of the real numbers $\mathbb{R}$:

$$\boldsymbol{X} := \boldsymbol{x}^{(1)} \times \boldsymbol{x}^{(2)} \times \ldots \times \boldsymbol{x}^{(n)}, \text{ with } \boldsymbol{x}^{(i)} = \left[\underline{x}^{(i)},\, \overline{x}^{(i)}\right] \text{ for } i \in \{1, 2, \ldots, n\}. \tag{3.8}$$

This notation can be interpreted as projection of the i-th interval component $\boldsymbol{x}^{(i)}$ to the i-th axis of the vector space. An illustration for $n = 2$ and $n = 3$ is given in Fig. 3.1.

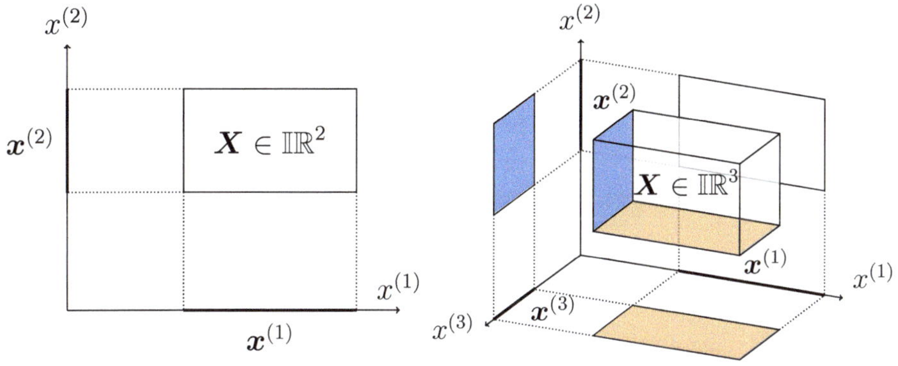

Figure 3.1: Examples of the graphical representation of $\boldsymbol{X} \in \mathbb{IR}^2$ (left) and $\boldsymbol{X} \in \mathbb{IR}^3$ (right)

An $(m \times n)$, $m, n \in \mathbb{N}$, interval matrix $\boldsymbol{A}$ can be interpreted as subspace of $\mathbb{R}^{m \times n}$. Again it is defined using the cartesian product of $m \cdot n$ closed intervals:

$$\begin{aligned} \boldsymbol{A} &= \begin{pmatrix} \boldsymbol{a}^{(1,1)} & \dots & \boldsymbol{a}^{(1,n)} \\ \vdots & & \vdots \\ \boldsymbol{a}^{(m,1)} & \dots & \boldsymbol{a}^{(m,n)} \end{pmatrix} \qquad (3.9) \\ &= \boldsymbol{a}^{(1,1)} \times \boldsymbol{a}^{(1,2)} \times \dots \times \boldsymbol{a}^{(m,n)} \\ &= \left(\boldsymbol{a}^{(i,j)}\right) \qquad (3.10) \end{aligned}$$

with $1 \leq i \leq m, 1 \leq j \leq n$. The center matrix is defined element-wise as in [Hla14] to

$$A_c \in \mathbb{R}^{m \times n} : \left(a_c^{(i,j)}\right) = \frac{1}{2}\left(\overline{a}^{(i,j)} + \underline{a}^{(i,j)}\right) \qquad (3.11)$$

as well as the radius matrix

$$A_\Delta \in \mathbb{R}^{m \times n} : \left(a_\Delta^{(i,j)}\right) = \frac{1}{2}\left(\overline{a}^{(i,j)} - \underline{a}^{(i,j)}\right) \qquad (3.12)$$

with $1 \leq i \leq m, 1 \leq j \leq n$.

The four basic arithmetic operations addition, subtraction, multiplication and division, i.e. $\star \in \{+, -, \cdot, /\}$, are well defined for intervals. The general application of each operator on two interval values $\boldsymbol{x} = [\underline{x}, \overline{x}]$ and $\boldsymbol{y} = [\underline{y}, \overline{y}]$ is given by

$$[\underline{x}, \overline{x}] \star [\underline{y}, \overline{y}] = \left\{z = x \star y \;\middle|\; \underline{x} \leq x \leq \overline{x}, \underline{y} \leq y \leq \overline{y}\right\} \qquad (3.13)$$

according to [Apo67], which leads to the interval type calculation rule

$$\begin{aligned} [\underline{x}, \overline{x}] \star [\underline{y}, \overline{y}] = \Big[&\min\left(\underline{x} \star \underline{y}, \underline{x} \star \overline{y}, \overline{x} \star \underline{y}, \overline{x} \star \overline{y}\right), \\ &\max\left(\underline{x} \star \underline{y}, \underline{x} \star \overline{y}, \overline{x} \star \underline{y}, \overline{x} \star \overline{y}\right) \Big]. \end{aligned} \qquad (3.14)$$

The different elements within the $\min(\cdot)$ and $\max(\cdot)$ operations are due to the fact that the combination of all extreme values need to be taken into account.

Unfortunately, this property also causes two major drawbacks of interval arithmetic, the dependency effect and the wrapping effect. Those effects are explained in Example 3.1 and Example 3.2.

With the assumption $0 \notin [\underline{y},\, \overline{y}]$ it is possible to explicitly state the four basic operations as in [Apo67]:

$$\begin{aligned}
[\underline{x},\, \overline{x}] + [\underline{y},\, \overline{y}] &= [\underline{x} + \underline{y},\, \overline{x} + \overline{y}] \\
[\underline{x},\, \overline{x}] - [\underline{y},\, \overline{y}] &= [\underline{x} - \overline{y},\, \overline{x} - \underline{y}] \\
[\underline{x},\, \overline{x}] \cdot [\underline{y},\, \overline{y}] &= [\min(\underline{x}\underline{y}, \underline{x}\overline{y}, \overline{x}\underline{y}, \overline{x}\overline{y}),\; \max(\underline{x}\underline{y}, \underline{x}\overline{y}, \overline{x}\underline{y}, \overline{x}\overline{y})] \\
[\underline{x},\, \overline{x}] \,/\, [\underline{y},\, \overline{y}] &= [\underline{x},\, \overline{x}] \cdot \left[\frac{1}{\overline{y}},\, \frac{1}{\underline{y}}\right].
\end{aligned} \tag{3.15}$$

It is shown in [Apo67] that associative and commutative property hold for interval values as well. However, the distributive law is not applicable anymore and needs to be changed to

$$\boldsymbol{x} \cdot (\boldsymbol{y} + \boldsymbol{z}) \subseteq \boldsymbol{x} \cdot \boldsymbol{y} + \boldsymbol{x} \cdot \boldsymbol{z} \tag{3.16}$$

which is known as the subdistributive property for the interval values $\boldsymbol{x}$, $\boldsymbol{y}$ and $\boldsymbol{z}$.

While evaluating an expression, every appearance of an interval variable is treated individually as if it was independent from its other occurrences. Multiple occurrences of the same variable thus lead to a widening of the enclosure. This property is called dependency effect and is illustrated in Example 3.1. One approach to mitigate the dependency effect is to reformulate the expression such that each variable occurs only once, if possible.

Example 3.1:

Assume the function

$$f(x) = 1 + \frac{1}{x}. \tag{3.17}$$

The resulting enclosure for the interval $\boldsymbol{x} = [1,\ 3]$ *can be calculated straight forward using the interval arithmetic definitions of (3.15) to* $f(\boldsymbol{x}) = \left[\frac{4}{3},\ 2\right]$*, which matches the true range of the function within the interval. It is depicted by the blue dotted frame in Fig. 3.2.*

If (3.17) is reformulated such that there are multiple occurences of x *e.g.*

$$\tilde{f}(x) = \frac{x+1}{x} \tag{3.18}$$

the interval arithmetic evaluation yields $\tilde{f}(\boldsymbol{x}) = \left[\frac{2}{3},\ 4\right]$*. It can be seen that this is a large overestimation of the true values of the function within* $\boldsymbol{x}$*, depicted by the dashed frame in Fig. 3.2.*

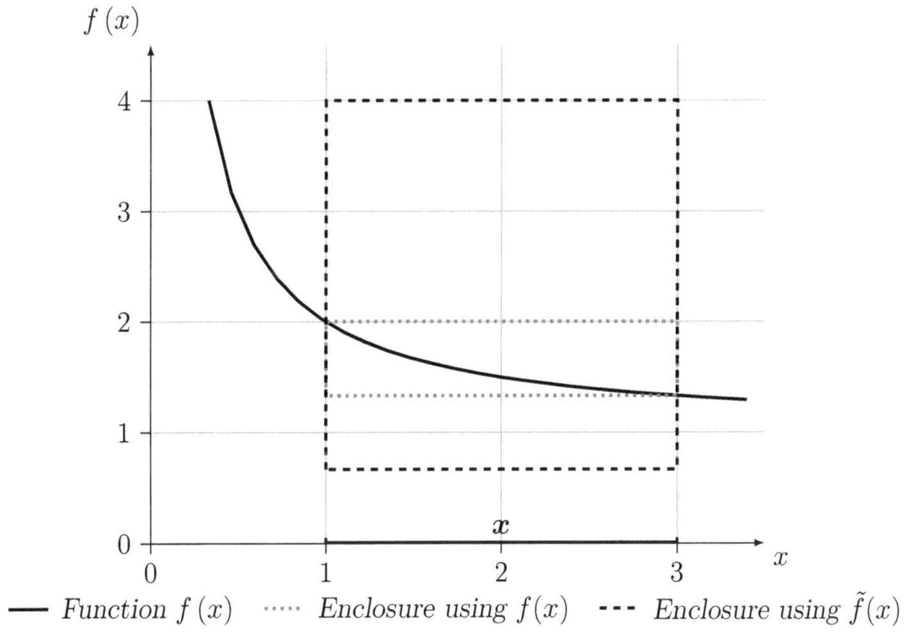

Figure 3.2: Dependency effect based on $f(x)$ and $\tilde{f}(x)$

Another effect occurring with interval calculations is the wrapping effect. This effect is caused by iterative calculations based on previous overestimations. Such iterative calculations are e.g. necessary to solve an initial value problem or to evaluate a state space equation. An example for the wrapping effect is given by the initial value problem of Moore [Bau87] and is depicted in Example 3.2.

Example 3.2:

Assume the initial value problem for the differential equation

$$\dot{\boldsymbol{Y}}(t) = \begin{pmatrix} 0 & 1 \\ -1 & 0 \end{pmatrix} \boldsymbol{Y}(t) \tag{3.19}$$

with $\boldsymbol{Y}(0) = [[-\epsilon,\ \epsilon],\ [1-\epsilon,\ 1+\epsilon]]^T$, $\epsilon > 0$. *When the equation is evaluated, the resulting solution set needs to be framed by axis parallel enclosures after each step. The solution sets for* $t_i = i \cdot \Delta t$, $\Delta t = \frac{\pi}{6}$, $i \in \{0, 1, 2, 3\}$ *are depicted in Fig. 3.3. It can be seen, that the overestimation is continually increasing, as the inherited overestimation is passed on and used as base for further calculations.*

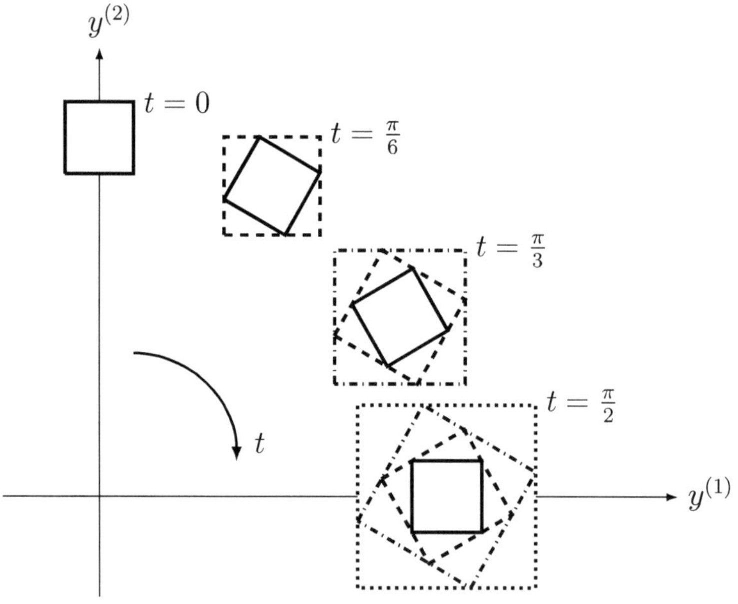

Figure 3.3: Example for the wrapping effect using the initial value problem of Moore [Bau87]

The wrapping effect can reach a serious extent even after only one iteration. A minimal example illustrating the extent of the problem after two steps is given in Example 3.3.

Example 3.3:

This example clarifies the effect of interval calculations in the setting of a proportional gain system with unknown gain p:

$$u \cdot p = y. \tag{3.20}$$

The system setup is depicted in Fig. 3.4. The input and output ranges of the system are given and can be included in the intervals $\boldsymbol{u} = [2,\ 3]$ *and* $\boldsymbol{y} = [4,\ 9]$. *The goal is to calculate the gain that maps all possible input values* $u \in \boldsymbol{u}$ *to the specified output range* $\boldsymbol{y}$.

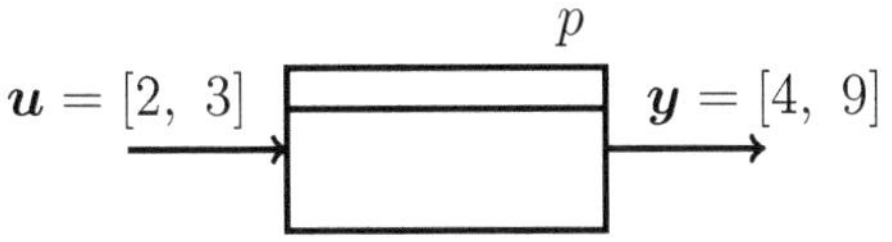

Figure 3.4: Proportional gain system with proper solution

Using the introduced interval arithmetic calculations leads to

$$\begin{aligned} \boldsymbol{p} &= \boldsymbol{y}/\boldsymbol{u} \\ &= [\underline{y},\ \overline{y}] \left[\frac{1}{\overline{u}},\ \frac{1}{\underline{u}}\right] \\ &= \left[\frac{4}{3},\ \frac{9}{2}\right] \\ &\approx [1.3,\ 4.5]\,. \end{aligned} \tag{3.21}$$

Re-substituting $\boldsymbol{p}$ *into the system equation yields*

$$\begin{aligned} \boldsymbol{y} &= \boldsymbol{p}\boldsymbol{u} \\ &= [1.3,\ 4.5]\,[2,\ 3] \\ &= [2.6,\ 13.5] \neq [4,\ 9] \end{aligned} \tag{3.22}$$

which is a strong overestimation of the genuine output range.

The example shows that the system parameter calculated from input and output ranges cannot be used to reason about the parameter set that is suitable to map the given input on the given output. The wrapping effect is caused by considering the combination of the extreme values of both intervals. Nevertheless, when regarding the task at hand in Example 3.3, the goal is not to find all possible gains connecting the two intervals but to find those gains reasonably connecting "the most" elements of the intervals. This slight but very important change is illustrated in Example 3.4.

Example 3.4:

Assume the setting of Example 3.3. The parameter interval is calculated as before using $\boldsymbol{p} = \boldsymbol{y}/\boldsymbol{u}$. *The question is now how many pairs* $(u, y)|(u \in \boldsymbol{u}), (y \in \boldsymbol{y})$ *exist for each parameter* $p \in \boldsymbol{p}$. *Therefore the intervals* $\boldsymbol{u}$ *and* $\boldsymbol{y}$ *are divided into equidistant parts of* $\Delta u = \Delta y = 0.0001$. *The resulting* $10'001$ *discrete samples of* u_s *are combined with the resulting* $50'001$ *samples of* y_s *to calculate the connecting parameter* $p_s = y_s/u_s$. *The histogram formed by* $500'060'001$ *values of* p_s *is depicted in Fig. 3.5. It can be seen that the extreme values of the outer enclosure of the solution* $\boldsymbol{p} = [1.3,\ 4.5]$ *are only connected by a single input-output pair each. On the other hand, there is a plateau between* $\boldsymbol{p}_i = [2,\ 3]$ *that connects a nearly constant number of input-output pairs. Substituting this interval value into the system equation leads to*

$$\begin{aligned} \boldsymbol{y}_i &= \boldsymbol{p}_i \boldsymbol{u} \\ &= [2,\ 3]\,[2,\ 3] \\ &= [4,\ 9] = \boldsymbol{y} \end{aligned} \tag{3.23}$$

which is exactly the given output range. The interval $\boldsymbol{p}_i$ *is an inner enclosure of the solution set of* $\boldsymbol{p} = \boldsymbol{y}/\boldsymbol{u}$.

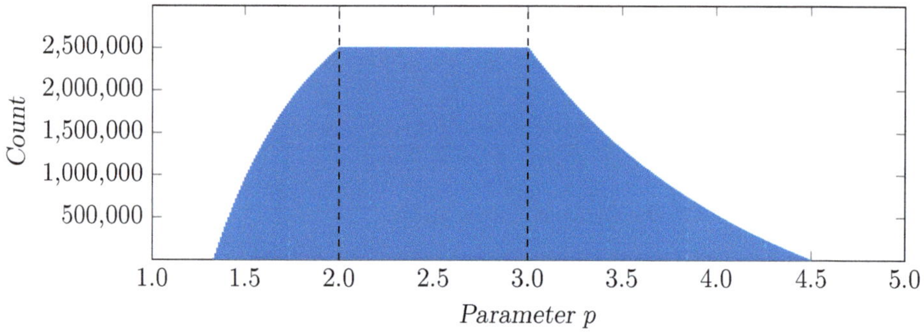

Figure 3.5: Distribution of parameters in the proper case

The plateau in Fig 3.5 contains those parameters $\boldsymbol{p}$ that are able to map any $u \in \boldsymbol{u}$ to a value $y \in \boldsymbol{y}$. Note that not necessarily all values $y \in \boldsymbol{y}$ have to be met by $\boldsymbol{p}\boldsymbol{u}$. The contour of the histogram given in Fig. 3.5 can also be analytically calculated. The derivation of the exact distribution is given in Appendix B.

The property leading to the wrapping effect displayed in Example 3.3 and Example 3.4 is the non-existence of an inverse element in classical interval arithmetic [Apo67].The inverse element in the real numbers is defined with respect to an operation and denotes an element that maps itself on the neutral element of this operation (see [Bro08, p. 340]).

Thereby the neutral element is also defined with respect to the same operation and denotes an element that maps each other element on itself (see [Bro08, p. 339]).
There are neutral elements in classical interval arithmetic. For example the neutral element for addition is given by $\boldsymbol{e}_a = [0,\ 0]$ and for multiplication by $\boldsymbol{e}_m = [1,\ 1]$. Applying the neutral elements to an arbitrary interval value $\boldsymbol{r} = [\underline{r},\ \overline{r}]$ leads to

$$\boldsymbol{r} + \boldsymbol{e}_a = [\underline{r} + 0,\ \overline{r} + 0] = [\underline{r},\ \overline{r}] \tag{3.24}$$

$$\boldsymbol{r} \cdot \boldsymbol{e}_m = [\underline{r} \cdot 1,\ \overline{r} \cdot 1] = [\underline{r},\ \overline{r}]\,. \tag{3.25}$$

However, in general there is no inverse element as can be seen in the following:

$$\boldsymbol{r} + (-\boldsymbol{r}) = [\underline{r} + (-\overline{r}),\ \overline{r} + (-\underline{r})] \neq \boldsymbol{e}_a \text{ if } \underline{r} \neq \overline{r} \tag{3.26}$$

$$\boldsymbol{r} \cdot \left(\frac{1}{\boldsymbol{r}}\right) = \left[\frac{\underline{r}}{\overline{r}},\ \frac{\overline{r}}{\underline{r}}\right] \neq \boldsymbol{e}_m \text{ if } \underline{r} \neq \overline{r}. \tag{3.27}$$

Equations (3.26) and (3.27) hold if and only if $\underline{r} = \overline{r}$ which means that $\boldsymbol{r}$ is a degenerated point real interval [Apo67].

3.1.2 Kaucher Interval Arithmetic

It is beneficial to define an extension to interval arithmetic that provides the existence of an inverse element for all arithmetic operations. By using Kaucher interval arithmetic [Kau80], the set of proper intervals can be extended by the introduction of a new set of so called improper intervals. These improper intervals are defined complementary to classical intervals:

$$\mathbb{KR} = \{\mathbf{x} = [\underline{x}, \overline{x}] \mid \overline{x} < \underline{x} \text{ and } \underline{x}, \overline{x} \in \mathbb{R}\}\,. \tag{3.28}$$

The set of all proper and improper intervals is given by $\mathbb{IR}^* = \mathbb{IR} \cup \mathbb{KR}$ and is depicted in Fig. 3.6. The set of point real intervals is depicted as diagonal line in the figure. The set of proper intervals $\mathbb{IR}$ is formed by the half plain above the point real line. It can be seen, that the improper intervals $\mathbb{KR}$ complete the $\mathbb{IR}^*$ by covering the half plain below the point real line.

The definitions of the basic arithmetic operations $\star \in \{+, -, \cdot, /\}$ need to be adapted to hold as well for classical as for Kaucher interval arithmetic [Sha02].
Therefore, the definition of the negative part $x^{\ominus}$ and the positive part $x^{\oplus}$ of a real number x is given by

$$x^{\ominus} = \max(-x, 0) \quad \text{and} \quad x^{\oplus} = \max(x, 0)\,. \tag{3.29}$$

The four classic operations can thus be written as follows:

$$\boldsymbol{x} + \boldsymbol{y} = [\underline{x} + \underline{y},\ \overline{x} + \overline{y}] \tag{3.30}$$

$$\boldsymbol{x} - \boldsymbol{y} = [\underline{x} - \overline{y},\ \overline{x} - \underline{y}] \tag{3.31}$$

$$\begin{aligned}\boldsymbol{x} \cdot \boldsymbol{y} = [&\max(\underline{x}^{\oplus}\underline{y}^{\oplus}, \overline{x}^{\ominus}\overline{y}^{\ominus}) - \max(\overline{x}^{\oplus}\underline{y}^{\ominus}, \underline{x}^{\ominus}\overline{y}^{\oplus}),\\ &\max(\underline{x}^{\ominus}\underline{y}^{\ominus}, \overline{x}^{\oplus}\overline{y}^{\oplus}) - \max(\overline{x}^{\ominus}\underline{y}^{\oplus}, \underline{x}^{\oplus}\overline{y}^{\ominus})]\end{aligned} \tag{3.32}$$

$$\boldsymbol{x}/\boldsymbol{y} = \boldsymbol{x} \cdot [1/\overline{y},\ 1/\underline{y}]\,, \text{for } \underline{y} \cdot \overline{y} > 0. \tag{3.33}$$

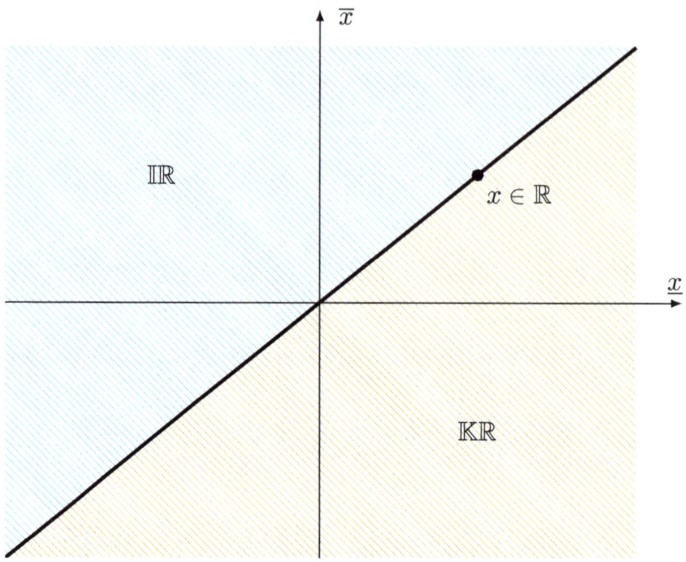

Figure 3.6: Geometric interpretation of $\mathbb{IR}^*$. The diagonal line represents point real values (based on [Sai14, p. 18])

In addition the two unary operators

$$\mathrm{opp}\left([\underline{x},\ \overline{x}]\right) = [-\underline{x},\ -\overline{x}] \tag{3.34}$$

$$\mathrm{dual}\left([\underline{x},\ \overline{x}]\right) = [\overline{x},\ \underline{x}]\,. \tag{3.35}$$

are introduced to toggle between proper and improper intervals. Using this operators leads to the definition of inverse elements in Kaucher interval arithmetic:

$$\boldsymbol{x} + \mathrm{opp}\left(\boldsymbol{x}\right) = [\underline{x},\ \overline{x}] + [-\underline{x},\ -\overline{x}] = [0,\ 0] =: 0 \tag{3.36}$$

$$\boldsymbol{x}/\mathrm{dual}\left(\boldsymbol{x}\right) = [\underline{x},\ \overline{x}] \cdot [1/\underline{x},\ 1/\overline{x}] = [1,\ 1] =: 1. \tag{3.37}$$

Those comply with the classic interval analysis definitions if all used intervals are proper [Sha02].

It is hard to imagine the nature of an improper interval as it is neither empty nor does it include the same values as a proper interval with inverse borders. A possibility to grasp an idea of the nature of an improper interval is given in Example 3.5.

Example 3.5:

Assume the proportional gain setting of Fig. 3.7 which is similar to Example 3.3 but with a different output range $\boldsymbol{y}$.

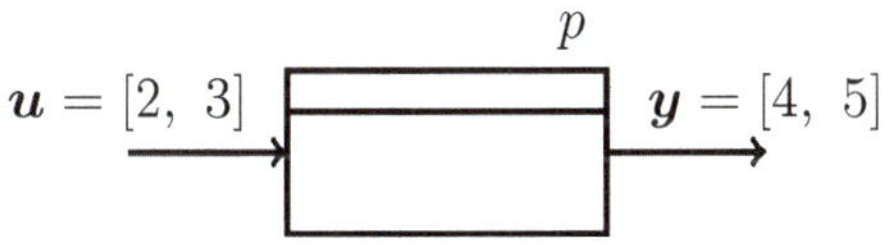

Figure 3.7: Proportional gain system with improper solution

The question is again which values can be used as gain $p \in \boldsymbol{p}$ that maps all input values $\boldsymbol{u} = [2,\ 3]$ to the output range $\boldsymbol{y} = [4,\ 5]$. The intervals are again divided into equidistant parts of $\Delta u = \Delta y = 0.0001$. The resulting $10'001$ discrete samples of u_s are combined with the $10'001$ samples of y_s to calculate the connecting parameter $p_s = y_s / u_s$. The resulting $100'020'002$ parameter values are used to set up the histogram given in Fig. 3.8.

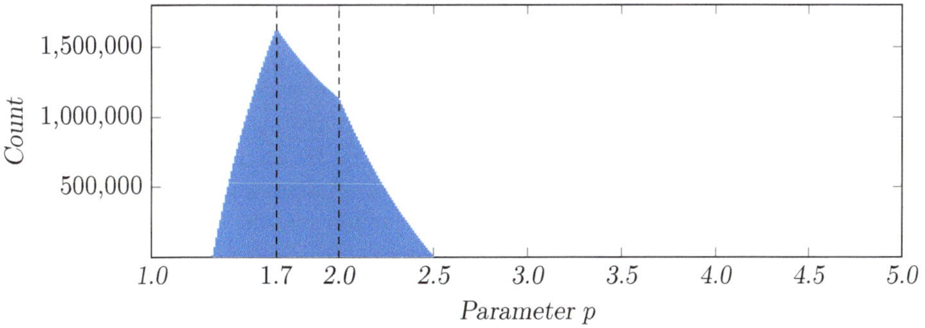

Figure 3.8: Distribution of parameters in the improper case

In this case it can be seen that there is no plateau in the histogram. However, there are two edges at $p_1 = 5/3$ and $p_2 = 2$. Substituting p_1 into the system equation leads to

$$\begin{aligned} \boldsymbol{y}_{p_1} &= p_1 \boldsymbol{u} \\ &= 5/3\,[2,\ 3] \\ &\approx [3.33,\ 5] \neq [4,\ 5]\,. \end{aligned} \tag{3.38}$$

The evaluation for p_2 yields

$$\begin{aligned} \boldsymbol{y}_{p_2} &= p_2 \boldsymbol{u} \\ &= 2\,[2,\ 3] \\ &\approx [4,\ 6] \neq [4,\ 5]\,. \end{aligned} \tag{3.39}$$

It can be seen that both parameters are able to map some values of the input range $\boldsymbol{u}$ *into the output range. Nevertheless there is not a single parameter that can map all values of* $\boldsymbol{u}$ *to* $\boldsymbol{y}$*. This observation can be combined with the fact that the inner enclosure is an improper interval:*

$$\begin{aligned} \boldsymbol{p} &= \boldsymbol{y}/dual(\boldsymbol{u}) \\ &= [\underline{y},\, \overline{y}] \left[\frac{1}{\underline{u}},\, \frac{1}{\overline{u}}\right] \\ &= \left[\frac{4}{2},\, \frac{5}{3}\right] \\ &\approx [2,\, 1.7] \end{aligned} \tag{3.40}$$

Therefore improper intervals can be interpreted as solutions of a setting with "eroded plateau".

3.1.3 Interval Type Linear Equation Systems

The introduced interval arithmetic considerations can now be extended to a vector matrix notation. Assume there are T measurement values $\langle \boldsymbol{u}_k \rangle_{k=1}^{T} = [\boldsymbol{u}_1, \boldsymbol{u}_2, \ldots, \boldsymbol{u}_T]^T$ and $\langle \boldsymbol{y}_k \rangle_{k=1}^{T} = [\boldsymbol{y}_1, \boldsymbol{y}_2, \ldots, \boldsymbol{y}_T]^T$, containing a valid range for each sample $k \in \{1, 2, \ldots, T\}$. Each suitable parameter $p \in \boldsymbol{p}$ has to comply with all input and all output ranges. This problem can be stated as an interval type linear equation system

$$\begin{cases} \boldsymbol{u}_1 \boldsymbol{p} &=& \boldsymbol{y}_1 \\ \boldsymbol{u}_2 \boldsymbol{p} &=& \boldsymbol{y}_2 \\ \vdots &=& \vdots \\ \boldsymbol{u}_T \boldsymbol{p} &=& \boldsymbol{y}_T \end{cases} \tag{3.41}$$

or more general, for vectorial input $\left[\boldsymbol{u}_k^{(1)}, \cdots, \boldsymbol{u}_k^{(n)}\right]$, scalar output $\boldsymbol{y}_k$ and n parameters $\left[\boldsymbol{p}^{(1)}, \cdots, \boldsymbol{p}^{(n)}\right]^T$:

$$\begin{cases} \boldsymbol{u}_1^{(1)} \boldsymbol{p}^{(1)} &+& \boldsymbol{u}_1^{(2)} \boldsymbol{p}^{(2)} &+& \ldots &+& \boldsymbol{u}_1^{(n)} \boldsymbol{p}^{(n)} &=& \boldsymbol{y}_1 \\ \boldsymbol{u}_2^{(1)} \boldsymbol{p}^{(1)} &+& \boldsymbol{u}_2^{(2)} \boldsymbol{p}^{(2)} &+& \ldots &+& \boldsymbol{u}_2^{(n)} \boldsymbol{p}^{(n)} &=& \boldsymbol{y}_2 \\ \vdots && \vdots &&&& \vdots && \vdots \\ \boldsymbol{u}_T^{(1)} \boldsymbol{p}^{(1)} &+& \boldsymbol{u}_T^{(2)} \boldsymbol{p}^{(2)} &+& \ldots &+& \boldsymbol{u}_T^{(n)} \boldsymbol{p}^{(n)} &=& \boldsymbol{y}_T. \end{cases} \tag{3.42}$$

The variables are used to set up the regressor matrix $\boldsymbol{A} \in \mathbb{IR}^{(T \times n)}$, the measurement vector $\boldsymbol{B} \in \mathbb{IR}^{(T \times 1)}$ and the respective parameter vector $\boldsymbol{X} \in \mathbb{IR}^{*(n \times 1)}$ with

$$\boldsymbol{A} = \left(\boldsymbol{a}^{(i,j)}\right)_{1 \leq i \leq T,\ 1 \leq j \leq n} = \left(\boldsymbol{u}_k^{(j)}\right)_{1 \leq k \leq T,\ 1 \leq j \leq n} \tag{3.43}$$

$$\boldsymbol{B} = \left(\boldsymbol{b}^{(i)}\right)_{1 \leq i \leq T} = \left(\boldsymbol{y}_k\right)_{1 \leq k \leq T} \tag{3.44}$$

$$\boldsymbol{X} = \left(\boldsymbol{x}^{(j)}\right)_{1 \leq j \leq n} = \left(\boldsymbol{p}^{(j)}\right)_{1 \leq j \leq n} \tag{3.45}$$

The interval type linear equation system can thus be stated as

$$\boldsymbol{A} \cdot \boldsymbol{X} = \boldsymbol{B}. \tag{3.46}$$

This system can be interpreted as the collection of all point real linear equation systems that can be formed from the enclosed interval values [Sha96].
Equation systems with a regressor matrix of dimension $T = n$ are called quadratic. Dimension $T < n$ stands for an underdetermined and dimension $T > n$ results in an overdetermined equation system. Underdetermined systems do not carry enough information to solve the problem unambigiously. This thesis focuses on overdetermined systems which is the most relevant case when regarding reasonable measurement times T and system orders n.

The inverse of a quadratic point real matrix A is defined if the matrix is non-singular i.e. A^{-1} exists if $\det(A) \neq 0$. Analogously a quadratic interval type matrix $\boldsymbol{A}$ is non-singular if all point real matrices contained in the interval matrix are non-singular i.e. $\det(A) \neq 0\, \forall A \in \boldsymbol{A}$ [Sha14].
For overdetermined point real systems the criterion changes to a rank condition. The point real matrix $A \in \mathbb{R}^{(T \times n)}$ with $T > n$ is said to have full rank if rank$\,(A) = n$. For interval type overdetermined systems, this condition again changes to rank$\,(A) = n$, $\forall A \in \boldsymbol{A}$. This means that all point real matrices included in the interval matrix need to show full rank.
Determining the rank of an interval type matrix is in general an NP-Hard problem [Sha14]. Nevertheless, several criteria to check if an interval matrix has full rank were collected in [Sta16] based on [Sha14]. An introduction of the most relevant ones is given in Appendix C.

The full rank condition is connected with persistent excitation according to Assumption 2.1. If the used input signal provides persistent excitation, the regressor matrix has full rank [Sha14][Lak14]. This leads to Assumption 3.1.

Assumption 3.1 (Rank of the Regressor Matrix)
The interval type regressor matrix $\boldsymbol{A}$ *shows full rank according to [Sha96].*

Throughout this thesis it is assumed that all specifications and measurements lead to an interval regressor matrix $\boldsymbol{A}$ that has full rank.

In general there is no unique, component wise point real solution vector X for such an interval linear equation system. Instead, (3.46) is solved by a set of point real solutions $\sum$. The elements of the solution set $X^{(i)} \in \sum$ depend on the specific interpretation of (3.46). This interpretation is done by the interval quantors $\forall$ and $\exists$ as explained in [Sha02]. The notation of

$$\forall \left[\underline{a},\, \overline{a}\right] x = \exists \left[\underline{b},\, \overline{b}\right] \tag{3.47}$$

means that x has to solve (3.47) for all elements of $\{a \in \mathbb{R} | a \in [\underline{a},\, \overline{a}]\}$ but only for at least one specific element of $\left\{b \in \mathbb{R} | b \in \left[\underline{b},\, \overline{b}\right]\right\}$ [Sha02].
Each element of a vector or matrix can be assigned with an individual quantor, i.e. it is possible to precisely define a specific solution set for the interval type matrix equations. Vectors and matrices with assigned quantors are denoted as $\boldsymbol{B}^{\mathfrak{C}}$ or $\boldsymbol{A}^{\mathfrak{C}}$, respectively. A vector or matrix containing only the elements with assigned $\forall$ quantor are denoted by $\boldsymbol{B}^{\forall}$ and $\boldsymbol{A}^{\forall}$, the elements assigned with an $\exists$ quantor are given by $\boldsymbol{B}^{\exists}$ and $\boldsymbol{A}^{\exists}$. To split an assigned vector $\boldsymbol{B}^{\mathfrak{C}}$ or a matrix $\boldsymbol{A}^{\mathfrak{C}}$ depending on the quantors, the dualization of the intervals as given in (3.37) has to be used.
According to [Sha02] the splitting is different for matrices and vectors and it is given by the following relation:

$$\text{Vector: } \boldsymbol{B}^{\mathfrak{C}} := \text{dual}\left(\boldsymbol{B}^{\forall}\right) + \boldsymbol{B}^{\exists} \tag{3.48}$$

$$\text{Matrix: } \boldsymbol{A}^{\mathfrak{C}} := \boldsymbol{A}^{\forall} + \text{dual}\left(\boldsymbol{A}^{\exists}\right). \tag{3.49}$$

The specific elements of the matrices $\boldsymbol{A}^{\forall}$ and $\boldsymbol{A}^{\exists}$ are given by

$$\boldsymbol{a}^{\forall(i,j)} = \begin{cases} \boldsymbol{a}^{\mathfrak{C}(i,j)} & \text{, if } \mathfrak{C} = \forall \\ 0 & \text{, else} \end{cases} \tag{3.50a}$$

$$\boldsymbol{a}^{\exists(i,j)} = \begin{cases} \boldsymbol{a}^{\mathfrak{C}(i,j)} & \text{, if } \mathfrak{C} = \exists \\ 0 & \text{, else.} \end{cases} \tag{3.50b}$$

The elements for the vectors $\boldsymbol{B}^{\forall}$ and $\boldsymbol{B}^{\exists}$ are given by

$$\boldsymbol{b}^{\forall(i)} = \begin{cases} \boldsymbol{b}^{\mathfrak{C}(i)} & \text{, if } \mathfrak{C} = \forall \\ 0 & \text{, else} \end{cases} \tag{3.51a}$$

$$\boldsymbol{b}^{\exists(i)} = \begin{cases} \boldsymbol{b}^{\mathfrak{C}(i)} & \text{, if } \mathfrak{C} = \exists \\ 0 & \text{, else.} \end{cases} \tag{3.51b}$$

The most general solution set is given by a mixed assignment of both quantors to the interval matrix $\boldsymbol{A}$ as well as to the interval vector $\boldsymbol{B}$. The resulting AE-solution[3] set $\sum_{AE}$ according to [Hla14] is given by

$$\begin{aligned} &\textstyle\sum_{AE}\left(\boldsymbol{A}^{\mathfrak{C}}, \boldsymbol{B}^{\mathfrak{C}}\right) = \{X \in \mathbb{R}^n | \\ &\left(\forall A^{\forall} \in \boldsymbol{A}^{\forall}\right) \wedge \left(\forall B^{\forall} \in \boldsymbol{B}^{\forall}\right) \wedge \left(\exists A^{\exists} \in \boldsymbol{A}^{\exists}\right) \wedge \left(\exists B^{\exists} \in \boldsymbol{B}^{\exists}\right) : \left(\left(A^{\forall} + A^{\exists}\right) X = B^{\forall} + B^{\exists}\right)\}. \end{aligned} \tag{3.52}$$

3 The genuine notation of [Hla14, p. 2] is:
$X \in \mathbb{R}^n$ is an AE-solution if $\forall A^{\forall} \in \boldsymbol{A}^{\forall}, \forall B^{\forall} \in \boldsymbol{B}^{\forall}, \exists A^{\exists} \in \boldsymbol{A}^{\exists}, \exists B^{\exists} \in \boldsymbol{B}^{\exists} : (A^{\forall} + A^{\exists})X = B^{\forall} + B^{\exists}$.
This notation is slightly adapted for the sake of readability.

Based on the general AE-solution it is possible to define four distinct solution sets as given in [Sha96][Fie06][Hla14].

Definition 3.1 (United Solution Set $\sum_{\exists\exists}$)
The united solution set is formed by all solutions of any of the point real systems $A \cdot X = B$ with $A \in \boldsymbol{A}$ and $B \in \boldsymbol{B}$ that are included in the interval system:

$$\textstyle\sum_{\exists\exists} (\boldsymbol{A}, \boldsymbol{B}) := \{X \in \mathbb{R}^n | \, (\exists A \in \boldsymbol{A}) \wedge (\exists B \in \boldsymbol{B}) : (A \cdot X = B)\} \, . \tag{3.53}$$

Note that not all $A \in \boldsymbol{A}$ can match any element $B \in \boldsymbol{B}$ by multiplication with any $X \in \sum_{\exists\exists}$, and that not all $B \in \boldsymbol{B}$ can be calculated using any $X \in \sum_{\exists\exists}$ and all available $A \in \boldsymbol{A}$. Furthermore, the set $\sum_{\exists\exists}$ is not necessarily connected and not necessarily constrained by borders parallel to the coordinate axes. Using an enclosing interval $\boldsymbol{X} \supseteq \sum_{\exists\exists}$ will likely create spurious solutions.

Definition 3.2 (Tolerable Solution Set $\sum_{\forall\exists}$)
The tolerable solution set includes all values of X that solve the interval type linear equation system regardless of the chosen point real matrix $A \in \boldsymbol{A}$. This means the solution holds for all included point real matrices:

$$\textstyle\sum_{\forall\exists} (\boldsymbol{A}, \boldsymbol{B}) := \{X \in \mathbb{R}^n | \, (\forall A \in \boldsymbol{A}) \wedge (\exists B \in \boldsymbol{B}) : (A \cdot X = B)\} \, . \tag{3.54}$$

Note that not all $B \in \boldsymbol{B}$ can be calculated using any $X \in \sum_{\forall\exists}$ and all available $A \in \boldsymbol{A}$. The set $\sum_{\forall\exists}$ is not necessarily connected and not necessarily constrained by borders parallel to the coordinate axes. Using an enclosing interval $\boldsymbol{X} \supseteq \sum_{\forall\exists}$ will likely create spurious solutions.
The controllable solution set applies the same principle to the measurement vector $\boldsymbol{B}$.

Definition 3.3 (Controllable Solution Set $\sum_{\exists\forall}$)
The elements of the controllable solution set are feasible regardless of the chosen point real measurement vector $B \in \boldsymbol{B}$. This means there is a suitable regressor matrix $A \in \boldsymbol{A}$ for all possible point real measurement vectors:

$$\textstyle\sum_{\exists\forall} (\boldsymbol{A}, \boldsymbol{B}) := \{X \in \mathbb{R}^n | \, (\exists A \in \boldsymbol{A}) \wedge (\forall B \in \boldsymbol{B}) : (A \cdot X = B)\} \, . \tag{3.55}$$

Note that not all $A \in \boldsymbol{A}$ can match an element $B \in \boldsymbol{B}$ by multiplication with any $X \in \sum_{\exists\forall}$. The set $\sum_{\exists\forall}$ is not necessarily connected and not necessarily constrained by borders parallel to the coordinate axes. Using an enclosing interval $\boldsymbol{X} \supseteq \sum_{\exists\forall}$ will likely create spurious solutions.

A very strict criterion is given by the strong solution set.

Definition 3.4 (Strong Solution set $\sum_{\forall\forall}$)
The strong solution set includes only parameters X that solve the interval type linear equation system for any regressor matrix and any measurement vector:

$$\sum_{\forall\forall}(\boldsymbol{A},\boldsymbol{B}) := \{X \in \mathbb{R}^n | (\forall A \in \boldsymbol{A}) \wedge (\forall B \in \boldsymbol{B}) : (A \cdot X = B)\}. \tag{3.56}$$

None of the limitations of the previous solution sets is necessary for the strong solution. Nevertheless, the set $\sum_{\forall\forall}$ is not necessarily connected and not necessarily constrained by borders parallel to the coordinate axes. Using an enclosing interval $\boldsymbol{X} \supseteq \sum_{\forall\forall}$ will likely create spurious solutions.
The algebraic solution differs in its definition as it is not quantor based.

Definition 3.5 (Algebraic Solution Set $\sum_a$)
The algebraic solution is defined by the interval type vectors $\boldsymbol{X}_a$ that solve the interval type linear equation system straight forward:

$$\sum_a(\boldsymbol{A},\boldsymbol{B}) := \{\boldsymbol{X}_a \in \mathbb{IR}^n | (\boldsymbol{A} \cdot \boldsymbol{X}_a = \boldsymbol{B})\}. \tag{3.57}$$

Even though the elements of the algebraic solution are constrained parallel to the coordinate axes, the solution is ambiguous, i.e. there might be several or none solution vectors $\boldsymbol{X}_a$ that fulfill the equation [Kup95].
The different solution sets are related as they are subsets of each other. The united solution set is a superset of the algebraic solution set [Kup95], as well as of the tolerable and the controllable solution set [Sha96]

$$\sum_a(\boldsymbol{A},\boldsymbol{B}) \subseteq \sum_{\exists\exists}(\boldsymbol{A},\boldsymbol{B}) \tag{3.58}$$
$$\sum_{\forall\exists}(\boldsymbol{A},\boldsymbol{B}) \subseteq \sum_{\exists\exists}(\boldsymbol{A},\boldsymbol{B}) \tag{3.59}$$
$$\sum_{\exists\forall}(\boldsymbol{A},\boldsymbol{B}) \subseteq \sum_{\exists\exists}(\boldsymbol{A},\boldsymbol{B}). \tag{3.60}$$

The strong solution set on the other hand is a subset of the tolerable as well as of the controllable solution set [Fie06]

$$\sum_{\forall\forall}(\boldsymbol{A},\boldsymbol{B}) \subseteq \sum_{\forall\exists}(\boldsymbol{A},\boldsymbol{B}) \tag{3.61}$$
$$\sum_{\forall\forall}(\boldsymbol{A},\boldsymbol{B}) \subseteq \sum_{\exists\forall}(\boldsymbol{A},\boldsymbol{B}). \tag{3.62}$$

A visualization of the solution sets and their relations is given in Example 3.6.

Example 3.6:

Consider the following 2×2 *interval type linear equation system (ILES) taken from [Sha96]:*

$$\begin{pmatrix} [2,\ 4] & [-2,\ 1] \\ [-1,\ 2] & [2,\ 4] \end{pmatrix} \cdot \boldsymbol{X} = \begin{pmatrix} [-2,\ 2] \\ [-2,\ 2] \end{pmatrix}. \tag{3.63}$$

The controllable solution set $\sum_{\exists\forall}$ *and the strong solution set* $\sum_{\forall\forall}$ *are empty for the ILES (3.63). It can be seen in Fig. 3.9 that the algebraic solution* $\sum_a$ *is a subset of the tolerable solution set* $\sum_{\forall\exists}$ *which is a subset of the united solution* $\sum_{\exists\exists}$*. It is also clearly visible that neither the tolerable nor the united solution can be included in classical intervals parallel to the axes without creating spurious solutions.*

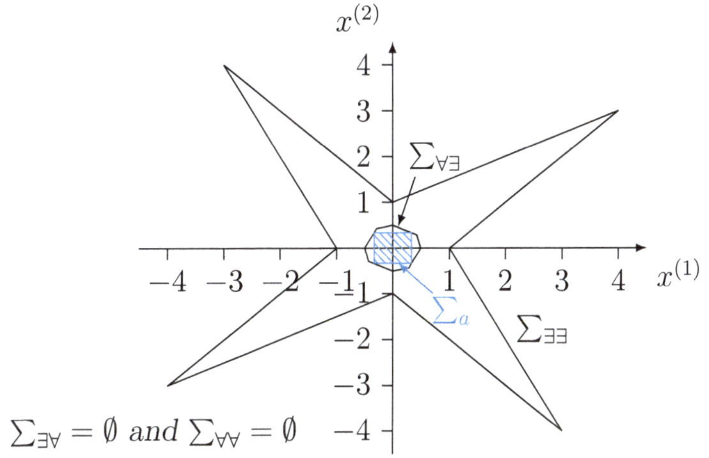

Figure 3.9: Different solution sets for the interval type linear equation system

The calculation of all given solution sets is computational expensive, as the calculation of the hulls is NP-Hard according to [Hor13]. Even to check whether a solution set is empty is still an NP-Complete problem as shown by [Sha96].

The problem becomes more tractable, if it is regarded from a different point of view. Assume a given point real solution candidate X_s. The question is now to determine whether the solution candidate X_s belongs to any of the defined solution sets without calculating the sets explicitly. The approach used in this thesis was introduced by [Bee72] and uses the so-called theorem of Prager-Oettli [Oet64]. The resulting criterion for interval arithmetic problems in Def. 3.6 is used to determine whether X_s is a member of the united solution set $\sum_{\exists\exists}$.

Definition 3.6 (Theorem of Prager-Oettli)
A given solution candidate X_s is part of the united solution set $\sum_{\exists\exists}$ i.e.

$$X_s \in \sum_{\exists\exists} \tag{3.64}$$

if and only if X_s fulfills the inequality

$$|A_c X_s - B_c| \leq A_\Delta |X_s| + B_\Delta \tag{3.65}$$

based on center and radius of the regressor matrix $\boldsymbol{A} = \langle A_c,\ A_\Delta \rangle$ and the measurement vector $\boldsymbol{B} = \langle B_c,\ B_\Delta \rangle$, given by the T interval type measurement values $\langle \boldsymbol{u}_k \rangle_{k=1}^{T}$ and $\langle \boldsymbol{y}_k \rangle_{k=1}^{T}$ [Bee72, p. 235].

This theorem was extended by [Hla14] to the general AE-solution:

$$X_s \in \sum_{AE} \Leftrightarrow |A_c X_s - B_c| \leq \left(A_\Delta^\exists - A_\Delta^\forall\right) |X_s| + B_\Delta^\exists - B_\Delta^\forall. \tag{3.66}$$

A solution candidate vector X_s is part of the AE-solution if and only if (3.66) holds. This criterion can be specialized to fit the four other solutions sets as given in Tab. 3.1.

Table 3.1: Conditions for the membership of X_s to a specific solution set.

Solution set	Condition
$\sum_{\exists\exists}$ (united)	$\lvert A_c X_s - B_c \rvert \leq A_\Delta \lvert X_s \rvert + B_\Delta$
$\sum_{\forall\exists}$ (tolerable)	$\lvert A_c X_s - B_c \rvert \leq -A_\Delta \lvert X_s \rvert + B_\Delta$
$\sum_{\exists\forall}$ (controllable)	$\lvert A_c X_s - B_c \rvert \leq A_\Delta \lvert X_s \rvert - B_\Delta$
$\sum_{\forall\forall}$ (strong)	$\lvert A_c X_s - B_c \rvert \leq -A_\Delta \lvert X_s \rvert - B_\Delta$

Further considerations regarding existence and uniqueness of the solution are only available for the algebraic solution set $\sum_a$. Two approaches for this purpose are sketched in Appendix D.

4 Guaranteed Verification of Point Real Systems

The theoretical foundation of the thesis is developed and illustrated in this chapter. Therefore a very simple and comprehensive linear time invariant model structure is used to focus on the method itself. The general principles introduced in this chapter can be extended to other types of system models.

4.1 System Setup

Definition 4.1 (Linear Time Invariant System)
A discrete time, linear time invariant (LTI) system can be modeled as

$$y_k = \sum_{i=1}^{n_a} a_i y_{k-i} + \sum_{i=1}^{n_c} c_i u_{k-i} \tag{4.1}$$

with the discrete time input u_k and output y_k, the input and output order n_a and n_c as well as the input parameters $[a_1, a_2, \ldots, a_{n_a}]^T$ and the output parameters $[c_1, c_2, \ldots, c_{n_c}]^T$. This modeling approach is also known as AutoRegressive system with eXogenous input *(ARX).*

Based on the model assumption of Def. 4.1 it is possible to set up the specification of the nominal system, as given in Def. 4.2. The set of nominal parameters as introduced in Sec. 2.1.1 is assumed to be given in the specification.[4] Two possibilities to determine these parameters in practice are introduced in Appendix E.
Throughout this thesis the superscript $\square^*$ will be used to denote values that are part of a specification or the nominal value. Note that the set of parameters is given by a distinguished point real vector for the current LTI setting.[5]

4 For other applications, e.g. fault-tolerant control, the method works similarly but the specification is given in a different manner.

5 The used model assumption does not allow a direct feedthrough as this property is not regarded in the given setting. To allow a direct feedthrough the second sum needs to be changed to start from zero, leading to $i \in \{0, 1, \ldots, n_c\}$.

Definition 4.2 (Specification of a Linear Time Invariant System)
The (direct) specification S_d^* of an LTI system according to Def. 4.1 is given by the mandatory values

- n_a^*, *the nominal output order*
- n_c^*, *the nominal input order*
- $\Theta^* = \left[a_1^*, a_2^*, \dots, a_{n_a^*}^*, c_1^*, c_2^*, \dots, c_{n_c^*}^*\right]^T$, *the nominal parameters*

and the optional values

- $Y_{init}^* = \langle y_k \rangle_{k=1}^{\max(n_a^*, n_c^*)}$, *the initial output values*
- $U_{init}^* = \langle u_k \rangle_{k=1}^{\max(n_a^*, n_c^*)}$, *the initial input values*

leading to the overall specification

$$S_d^* = \{\Theta^*, n_a^*, n_c^*, U_{init}^*, Y_{init}^*\} . \tag{4.2}$$

If the initial values are known, the future evolution of the system output trajectory is only dependent on the input signal. It is thus possible to compare the behavior of the trajectory for different inputs. If there are no initial values, the behavior of the trajectory will differ for the same inputs in the case of different used initial values. If they are provided, there need to be at least $k_{min} = \max(n_a^*, n_c^*) + 1$ initial values to enable the first evaluation of the autoregressive system description according to Def. 4.1.

It is assumed that the nominal system is developed and built and ready to be verified. Thereby the verification object (VO) is assumed to be available as a (physical) black box with one or more input and output ports. It is possible to excite the system via the input and to measure the resulting output. Further insights, like internal structure, components and wiring, software, plans or internal states are not accessible. This approach can be applied in various states of system development. Therefore there is a wide range of exact physical representations of the VO black box such as models, program code, components or units.

The verification method is running on a digital device that not necessarily generates the input signal itself. Therefore input and output values need to be measured to be available for the verification method. Measurement data is always subject to noise which is assumed to be modeled throughout the thesis based on the following definition.

Definition 4.3 (Sensor Noise Properties)
All available information about the VO is given in terms of measurement data of the input $\langle u_{meas,k}\rangle_{k=1}^{T}$ *and output* $\langle y_{meas,k}\rangle_{k=1}^{T}$. *The measurement data is obtained using sensors providing guaranteed notions of sensor precision that allow interval type enclosure of the measurement values.*
All measurement values $u_{meas,k}$ *and* $y_{meas,k}$ *are extended to intervals such that the true system values* $u_{true,k}$ *and* $y_{true,k}$ *are guaranteed to be included in the interval* $\boldsymbol{u}_{meas,k}$ *and* $\boldsymbol{y}_{meas,k}$, *respectively. There are three different ways these guarantees can be obtained:*

a) ***Absolute deviation***
The sensor precision is denoted by a maximum deviation of $\pm\delta^a$. *This leads to the interval enclosure of*

$$u_{true,k} \in \boldsymbol{u}_k = \left[u_{meas,k} - \delta_u^a,\ u_{meas,k} + \delta_u^a\right] \tag{4.3}$$

$$y_{true,k} \in \boldsymbol{y}_k = \left[y_{meas,k} - \delta_y^a,\ y_{meas,k} + \delta_y^a\right]. \tag{4.4}$$

b) ***Relative deviation***
In this case the sensor precision is given in terms of a relative deviation of $\delta^r \in [0,\ 1]$ *which is leading to*

$$u_{true,k} \in \boldsymbol{u}_k = \left[u_{meas,k} \cdot (1 - \delta_u^r),\ u_{meas,k} \cdot (1 + \delta_u^r)\right] \tag{4.5}$$

$$y_{true,k} \in \boldsymbol{y}_k = \left[y_{meas,k} \cdot (1 - \delta_y^r),\ y_{meas,k} \cdot (1 + \delta_y^r)\right]. \tag{4.6}$$

c) ***Combined deviation***
A common case is the combination of the both aforementioned deviation types. The deviation is defined to be $\delta^r \in [0,\ 1]$ *times the current measurement value but at least* $\pm\delta^a$, *resulting in*

$$u_{true,k} \in \boldsymbol{u}_k = u_{meas,k} \cdot \left[\min\left((1 - \delta_u^r), \left(1 - \frac{\delta_u^a}{u_{meas,k}}\right)\right),\right.$$
$$\left.\max\left((1 + \delta_u^r), \left(1 + \frac{\delta_u^a}{u_{meas,k}}\right)\right)\right] \tag{4.7}$$

$$y_{true,k} \in \boldsymbol{y}_k = y_{meas,k} \cdot \left[\min\left((1 - \delta_y^r), \left(1 - \frac{\delta_y^a}{y_{meas,k}}\right)\right),\right.$$
$$\left.\max\left((1 + \delta_y^r), \left(1 + \frac{\delta_y^a}{y_{meas,k}}\right)\right)\right]. \tag{4.8}$$

The properties of Def. 4.3 are used to set up interval type enclosures of the measurement data that are guaranteed to include the true system value. The resulting structural setup of the measurement process is depicted in Fig. 4.1.

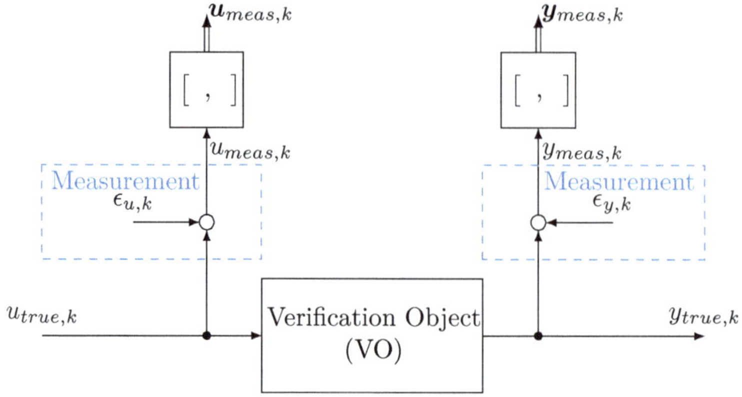

Figure 4.1: Structure of measurement setup

A basic assumption in the field of parameter identification is the property of persistent excitation according to Assumption 2.1. The information provided in any data set is highly dependent on the input signal that was used to generate the output. According to [Ast95, p. 63ff] there are several methods to ensure persistent excitation of a system. Exemplary persistently exciting inputs are e.g. white noise, pseudorandom binary sequences or a moving average process [Ise10, p. 251]. One possible excitation procedure fitted to the specific settings regarded in this thesis was developed in [Rie17]. The main idea is sketched in Appendix F.

4.2 Time Invariant Full Consistency

The general notion of consistency introduced in Chapter 2 is now transferred to the specific setting of LTI systems. The direct specification S_d^* includes one distinctive point real nominal parameter vector. Thus all results show *Full Consistency* according to Def. 2.2.
The observed behavior of the regarded VO is given in terms of input output measurement data. The nominal behavior is specified according to Def. 4.2. The VO is called full consistent with its specification if the measurement data can be explained by all specified parameters. The verification question is formally stated in Problem 4.1.

Problem 4.1 (Time Invariant Full Consistency)
Is the nominal system, specified by a direct specification

$$S_d^* = \{\Theta^*, n_a^*, n_c^*, U_{init}^*, Y_{init}^*\}, \tag{4.9}$$

full consistent with the input-output behavior given by the interval type enclosures of T measurement values

$$[\mathbf{U}_{meas}, \mathbf{Y}_{meas}] = \left[\langle \mathbf{u}_{meas,k}\rangle_{k=1}^{T}, \langle \mathbf{y}_{meas,k}\rangle_{k=1}^{T}\right] \tag{4.10}$$

i.e. can the measurement data be explained by the nominal system?

Problem 4.1 can be solved using the united solution set according to Def. 3.1.

Proposition 4.1 (Time Invariant Full Consistency)
The interval enclosure of the measurement data $[\mathbf{U}_{meas}, \mathbf{Y}_{meas}]$, given for the discrete sampling points $k = \{1, 2, \ldots, T\}$, leading to the interval type regressor matrix $\boldsymbol{A}_{meas}$ and the interval type measurement vector $\boldsymbol{B}_{meas}$, of a VO is called full consistent with a direct specification S_d^*, if the specified system parameters $\Theta^* = \left[a_1^*, a_2^*, \ldots, a_{n_a^*}^*, c_1^*, c_2^*, \ldots, c_{n_c^*}^*\right]^T$ are part of the united solution set $\sum_{\exists\exists}$ given by the measurement data, i.e.

$$\left(\Theta^* \in \textstyle\sum_{\exists\exists} (\boldsymbol{A}_{meas}, \boldsymbol{B}_{meas})\right) \Leftrightarrow Full\ Consistency^- \Rightarrow Full\ Consistency. \tag{4.11}$$

Proof:
The nominal parameter vector $\Theta^* = \left[a_1^*, a_2^*, \ldots, a_{n_a^*}^*, c_1^*, c_2^*, \ldots, c_{n_c^*}^*\right]^T$, given by a direct specification S_d^* according to Def. 4.2, can be interpreted as a solution candidate for the ILES (3.46) which is set up by the interval type enclosure of the measurement data. If Θ^* is part of the solution set of the ILES (3.46), the specification S_d^* is able to explain the measurement data.
The problem is formulated in Kaucher interval arithmetic, therefore it is necessary to define which solution set is used. Considering the interval enclosure of the measurement data given in Def. 4.3, it is obvious that it is not possible to determine the true value $u_{true,k}$ and $y_{true,k}$ as there are two distortion steps between the true values and the interval enclosure. First the true value is changed by the measurement random noise ϵ. Second, the sensor is only as precise as given by its property.

However, it is guaranteed that the true values of $u_{true,k}$ and $y_{true,k}$ are included in the measurement intervals

$$u_{true,k} \in \boldsymbol{u}_{meas,k} \tag{4.12}$$
$$y_{true,k} \in \boldsymbol{y}_{meas,k}. \tag{4.13}$$

Starting from time $k_{min} = \max(n_a^*, n_c^*) + 1$ there are enough measurement values to set up the regressor equations. Each additional measurement value leads to an additional row in the regressor matrix $\boldsymbol{A}$.
The unknown true values can be assumed to form an unknown true point real regressor matrix

$$A_{true} = \left[\begin{array}{ccc|ccc} y_{true,k_{min}-1} & \cdots & y_{true,k_{min}-n_a^*} & u_{true,k_{min}-1} & \cdots & u_{true,k_{min}-n_c^*} \\ y_{true,k_{min}} & \cdots & y_{true,k_{min}+1-n_a^*} & u_{true,k_{min}} & \cdots & u_{true,k_{min}+1-n_c^*} \\ \vdots & \ddots & \vdots & \vdots & \ddots & \vdots \\ y_{true,T-1} & \cdots & y_{true,T-1-n_a^*} & u_{true,T-1} & \cdots & u_{true,T-1-n_c^*} \end{array}\right] \tag{4.14}$$

and an unknown true point real measurement vector

$$B_{true} = \left[y_{true,k_{min}}, y_{true,k_{min}+1}, \ldots, y_{true,T}\right]^T. \tag{4.15}$$

These elements are linked via an unknown true parameter vector Θ_{true}, that fulfills

$$A_{true}\Theta_{true} = B_{true}. \tag{4.16}$$

Based on the enclosure of the true values in Def. 4.3 holds:

$$A_{true} \in \boldsymbol{A}_{meas} \tag{4.17}$$
$$B_{true} \in \boldsymbol{B}_{meas}. \tag{4.18}$$

With the set definition (3.53) follows that Θ_{true} is an element of the united solution set $\sum_{\exists\exists}(\boldsymbol{A}_{meas}, \boldsymbol{B}_{meas})$.
It is impossible to determine the true values A_{true} and B_{true} from the given measurement data. Therefore each point real element of the interval type regressor matrix and the interval type measurement vector is a possible true value. Thus the whole united solution set can be considered as correct solution of the ILES.
The given direct specification S_d^* shows $Full\ Consistency^-$ with the given measurement data $[\boldsymbol{U}_{meas}, \boldsymbol{Y}_{meas}]$, if and only if the nominal parameter vector Θ^* is part of the united solution set $\sum_{\exists\exists}(\boldsymbol{A}_{meas}, \boldsymbol{B}_{meas})$. Due to the underapproximating property holds

$$Full\ Consistency^- \Rightarrow Full\ Consistency \tag{4.19}$$

and thus the VO is full consistent in the sense of this thesis. □

The calculation of the united solution set is computationally expensive as introduced in Section 3.1.2. However, it is not necessary to calculate the whole solution set in this setting as there is a candidate solution given in form of the specification. Thus it is sufficient to check whether the specified parameter vector Θ^* is part of the united solution set without calculating the solution set explicitly. Prop. 4.1 can be checked very efficiently using the theorem of Prager-Oettli according to Definition 3.6 by evaluating the single equation (3.65). Therefore (3.65) is reformulated for the given measurement values to

$$|A_{meas,c}\Theta^* - B_{meas,c}| \leq A_{meas,\Delta} |\Theta^*| + B_{meas,\Delta} \Leftrightarrow \Theta^* \in \textstyle\sum_{\exists\exists} (\boldsymbol{A}_{meas}, \boldsymbol{B}_{meas}) . \tag{4.20}$$

As stated in Section 3.1.3 the existence and uniqueness of the solution sets is still an open question. A necessary condition for the existence of any solution set is that the ILES (3.46) is solvable. For the general overdetermined setting, this can be checked using the approaches given in Appendix C. However, their application is limited as the problem is in general NP-hard. The only further considerations regard the algebraic solution set and are sketched in Appendix D.

Note that the introduced method preserves time-invariance when checking for consistency. This is due to the used specification and represents a main difference to the direct image based methods used in fault detection as introduced in Section 2.2. This property will become even more clear in Chapter 5.

A further necessary condition is persistent excitation of the VO which is given in Assumption 2.1, developed to ensure full rank according to Assumption 3.1.
Therefore it is in general not guaranteed that there is a nonempty united solution set available and thus there are situation in which the proposed method is not applicable.
However, it is possible to facilitate a favorable situation by proper experiment design. One possibility to determine a beneficial excitation signal is given in Appendix F.

The application of time invariant full consistency for LTI systems is demonstrated in Example 4.1 and was presented to the scientific community in [Sch17b] and [Sch17c][Sch19].

Example 4.1:

This example shows the verification of a linear, time invariant system as introduced in Prop. 4.1. Assume the following direct specification

$$S_{d,1}^{*} = \{\Theta^{*} = [0.9825,\ 0.0675]\,, n_a^{*} = 1, n_c^{*} = 1, U_{init}^{*} = [0]\,, Y_{init}^{*} = [0]\}\,. \tag{4.21}$$

The simulations are done with a virtual VO, correctly implemented as discrete time linear ARX system

$$y_k = 0.9825 y_{k-1} + 0.0675 u_{k-1} \tag{4.22}$$

with sampling time $\Delta t = 1$s. The system is excited using a noise signal with uniformly distributed amplitude $u_{true,k} \in [0,\ 10]$ with mean $u_{mean} = 5$. It is assumed that the input is measured using a sensor with a maximum relative fault of $\delta_u^r = 0.05$. The resulting enclosure of the input measurement signal $\mathbf{U}_{meas}$ is depicted in the first subplot of Fig. 4.2. Nevertheless the system (4.22) is fed with the undisturbed input signal U_{true}. The resulting output signal is measured using a sensor with the same properties as the input sensor. The enclosed measurement output signal $\mathbf{Y}_{meas}$ is depicted in the second subplot of Fig. 4.2.

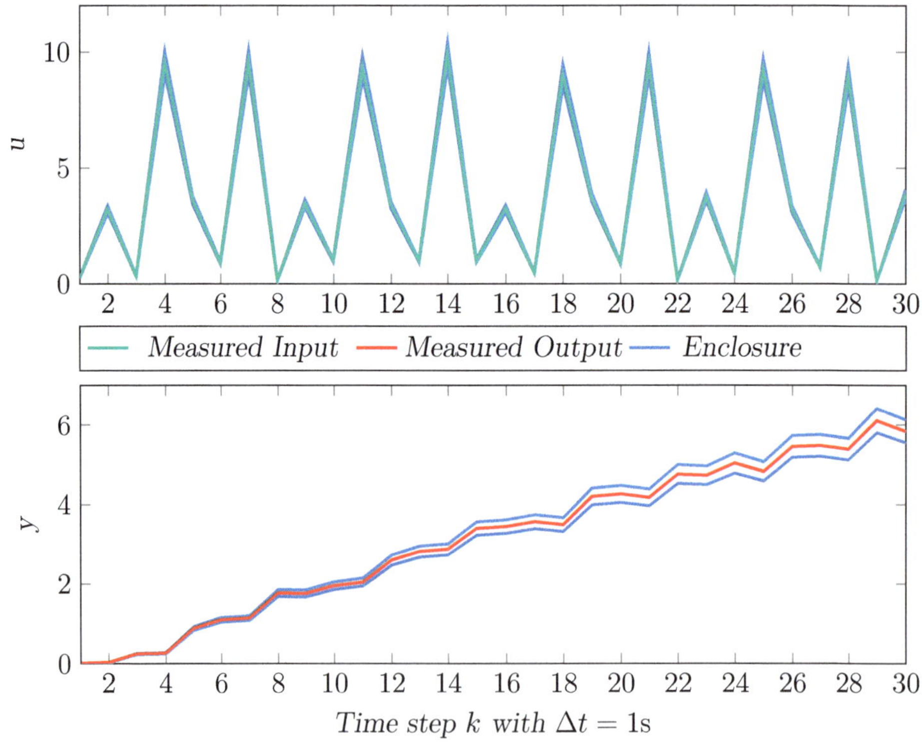

Figure 4.2: Measured input signal $\boldsymbol{U}_{meas}$ with $\delta_u^r = 0.05$

It can be shown that Prop. 4.1 holds for $S^*_{d,1}$ and the measurement data $\boldsymbol{Y}_{meas}$ and $\boldsymbol{U}_{meas}$ which proofs full consistency of measurement and specification formally.
A graphical representation is given in Fig. 4.3. The lines depict borders of the united solution set, generated by the different rows of the regressor matrix. Feasible parameters need to be located in between the borders of all rows of the measurement matrix. The parameters given in specification $S^*_{d,1}$ are marked with a green cross and form a feasible solution of the given problem as they are located within the united solution set of all measurement data. Hence it is possible to explain the measured data with the parameters given in specification $S^*_{d,1}$.
All given measurement values are used in this example to set up the regressor matrix and the measurement vector. This leads to $\boldsymbol{A}_{meas} \in \mathbb{IR}^{(29\times 2)}$. In this case, the full rank Assumption 3.1 for $n = 2$ parameters leads to rank$(\boldsymbol{A}_{meas}) \overset{!}{=} 2$ which is fulfilled for the given dynamic and excitation signal.
An example with failed verification can be given for the case that the verification method is applied using a different specification on the same measurement data. For this purpose

$$S^*_{d,2} = \{\Theta^* = [1.15,\ 0.08]\,, n^*_a = 1, n^*_c = 1, U^*_{init} = [0]\,, Y^*_{init} = [0]\}\,. \tag{4.23}$$

is used.

A graphical representation of the parameters is given by the red mark in Fig. 4.3. In this case, the parameters do not lie within the borders of the united solution set of the given measurement data. Thus the system is not verified.

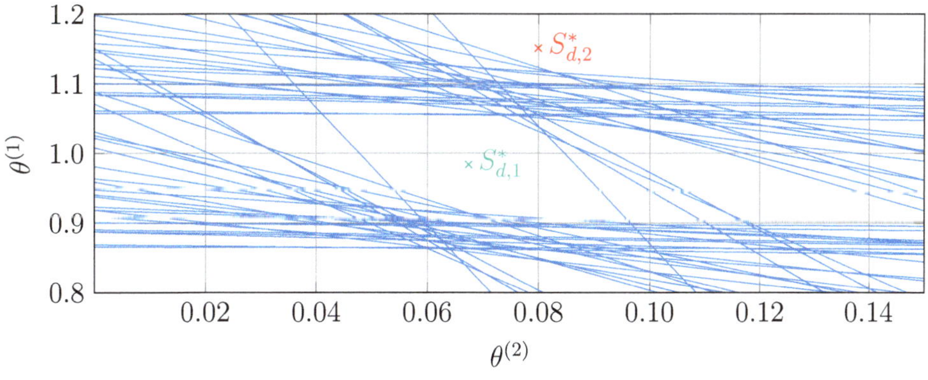

Figure 4.3: Visualization of the united solution given by the measurement data

4.3 Conclusion

The main idea of Kaucher arithmetic based verification was introduced in this chapter. Therefore the precise system and problem setting was defined and explained. The problem setup leads to a very distinct situation with only poor knowledge about the true measurement values which are enclosed in intervals. If this setting is regarded from a different point of view, it can be interpreted as a specific quantor based solution set definition that matches exactly the united solution set introduced in the mathematical preliminaries. This property can be used to check full consistency of the measurement data and the specification in a set membership procedure that is computationally very effective. The assumption of a full rank interval regressor matrix leads to preliminaries on the sensors and the noise assumptions. It is possible to check whether a specific solution candidate is part of the united solution given by measurement data. This property was demonstrated using an illustrative example.

The main advantage of the introduced method is that it focuses on the united solution set. Thereby it is possible to avoid wrapping and dependency effects and to calculate a solution set free of spurious solutions. This property is very beneficial in the case of safety critical systems as it avoids type II errors (hidden alarms).

5 Guaranteed Verification of Interval Type Systems

The basic idea introduced in the previous chapter is now extended to an interval type specification. Thus the parametrization is given by an interval type vector $\boldsymbol{\Theta}^*$ instead of a point real vector Θ^*.

Definition 5.1 (Interval Type Specification of a Linear System)
An interval type specification S_i^* of a linear system is given by the mandatory values

- n_a^*, *the nominal output order*
- n_c^*, *the nominal input order*
- $\boldsymbol{\Theta}^* = \left[\boldsymbol{a}_1^*, \boldsymbol{a}_2^*, \ldots, \boldsymbol{a}_{n_a^*}^*, \boldsymbol{c}_1^*, \boldsymbol{c}_2^*, \ldots, \boldsymbol{c}_{n_c^*}^*\right]$, *the interval type nominal system parameter vector*

and the optional values

- $Y_{init}^* = \langle y_k \rangle_{k=1}^{\max(n_a^*, n_c^*)}$, *the initial output values*
- $U_{init}^* = \langle u_k \rangle_{k=1}^{\max(n_a^*, n_c^*)}$, *the initial input values.*

This leads to the overall specification

$$S_i^* = \{\boldsymbol{\Theta}^*, n_a^*, n_c^*, U_{init}^*, Y_{init}^*\} \,. \tag{5.1}$$

Based on this specification the system is implemented. It is assumed that the resulting VO is given in a form that provides the input and output signals as described in the specification. Again, this can be the case for a variety of test objects, depending on the specific point of the development cycle for which the specification S_i^* was defined. Methods to determine the nominal parameters in practice are given in Appendix G.

Even though the specification is now given by interval type values, the implemented system has to provide real output data at any given time. Thus the real implementation of the VO has to use a specific real parametrization. This leads to the definition of interval type linear systems as given in Def. 5.2.

Definition 5.2 (Interval Type Linear System)
A discrete time, linear, interval type system can be modeled as

$$y_k = \sum_{i=1}^{n_a} a_{i,k} y_{k-i} + \sum_{i=1}^{n_c} c_{i,k} u_{k-i} \tag{5.2}$$

with the discrete time input u_k *and output* y_k*, the input and output order* n_a *and* n_c *as well as the time variant parameters*

$$\Theta_k = \left[a_{1,k}, a_{2,k}, \ldots a_{n_a,k}, c_{1,k}, c_{2,k}, \ldots c_{n_c,k}\right]^T \in \boldsymbol{\Theta}^*. \tag{5.3}$$

The necessary data is available as disturbed, discrete time measurement data enclosed in intervals according to Def. 4.3. Also the persistent excitation Assumption 2.1 and the full rank Assumptions 3.1 are still required to hold.

The given system definition leads to the time variant regressor vector

$$A_k = \left[y_{k-1}, y_{k-2}, \ldots, y_{k-n_a}, u_{k-1}, u_{k-2}, \ldots, u_{k-n_c}\right] \tag{5.4}$$

and thus the system equation can be transferred to

$$y_k = A_k \Theta_k \tag{5.5}$$

for a specific time step $k \geq k_{min}$ with $k_{min} = \max\left(n_a, n_c\right) + 1$.

This can be interpreted as the realization of a time variant system whose interval type specification is given according to Def. 5.3.

Definition 5.3 (Interval Enclosure of Time Variant Parameter)
The parameter values Θ_k *evolve during a specific time* $k \in \{1, 2, \ldots, T\}$ *and can be enclosed in the interval*

$$\Theta_k \in \boldsymbol{\Theta} = \left[\boldsymbol{\theta}^{(1)}, \boldsymbol{\theta}^{(2)}, \ldots, \boldsymbol{\theta}^{(n)}\right]^T \tag{5.6}$$

with $n = n_a + n_c$ *and* $\boldsymbol{\theta}^{(i)} = \left[\min\left(\left\langle \theta_k^{(i)} \right\rangle_{k=1}^{T}\right), \max\left(\left\langle \theta_k^{(i)} \right\rangle_{k=1}^{T}\right)\right]$*, denoting the minimum and maximum value of the* i*-th component within the regarded time.*

The time variance is given only in the parameters, the model structure, especially the model orders n_a and n_c, are time constant. Furthermore this interpretation is not necessarily beneficial for all time variant systems as the resulting interval enclosures can be very large depending on the time variant dynamic of the system parameters.

5.1 Interval Type Full Consistency

In the case of an interval type specification both consistency definitions ($Full\ Consistency$ and $Basic\ Consistency$) according to Def. 2.2 and Def. 2.1 are possible. In this section, the idea of $Full\ Consistency$ is extended to interval type systems. Then the situation is relaxed to $Basic\ Consistency$ in the next section.

Problem 5.1 (Interval Type Full Consistency)
Is the nominal system, specified by an interval type specification

$$S_i^* = \{\mathbf{\Theta}^*, n_a^*, n_c^*, U_{init}^*, Y_{init}^*\}, \tag{5.7}$$

full consistent with the input-output behavior given by the interval type enclosures of T *measurement values*

$$[\mathbf{U}_{meas}, \mathbf{Y}_{meas}] = \left[\langle \mathbf{u}_{meas,k}\rangle_{k=1}^{T}, \langle \mathbf{y}_{meas,k}\rangle_{k=1}^{T}\right] \tag{5.8}$$

i.e. do all elements of the parameter vector $\mathbf{\Theta}^*$ *fulfill Prop. 4.1 for all measurement data?*

The full consistency problem is solved by Prop. 5.1.

Proposition 5.1 (Interval Type Full Consistency)
The interval enclosure of the measurement data $[\mathbf{U}_{meas}, \mathbf{Y}_{meas}]$ *given for the discrete sampling points* $k = \{1, 2, \ldots, T\}$*, forming the regressor matrix* $\boldsymbol{A}_{meas}$ *and the measurement vector* $\boldsymbol{B}_{meas}$ *of a VO, is called to be full consistent with an interval type specification* S_i^**, if the complete set of specified parameters* $\mathbf{\Theta}^*$ *is part of the united solution set* $\sum_{\exists\exists}$ *given by the measurement data, i.e.*

$$\left(\mathbf{\Theta}^* \subseteq \sum\nolimits_{\exists\exists}(\boldsymbol{A}_{meas}, \boldsymbol{B}_{meas})\right) \Leftrightarrow Full\ Consistency^- \Rightarrow Full\ Consistency \tag{5.9}$$

Proof:
Full consistency follows directly from applying Prop. 4.1 to all possible point real parameter vectors $\Theta^* \in \mathbf{\Theta}^*$ given in the interval type specification S_i^*. □

The inverse relation of Prop. 5.1 leads to the implications given in Prop. 5.2.

> **Proposition 5.2 (Inverse of Full Consistency)**
> *If there is at least one $\Theta^* \in \mathbf{\Theta}^*$ that does not show full consistency according to Prop. 4.1, the interval type specification S_i^* is not full consistent with the measurement data.*

Proof:
Full consistency according to Prop. 5.1 is defined for all parameters $\Theta^* \in \mathbf{\Theta}^*$. A parameter $\tilde{\Theta}^* \in \mathbf{\Theta}^*$ that does not show full consistency according to Prop. 4.1 leads to

$$\mathbf{\Theta}^* \not\subseteq \sum\nolimits_{\exists\exists} (\boldsymbol{A}_{meas}, \boldsymbol{B}_{meas}) . \tag{5.10}$$

and thus the interval type specification S_i^* is not full consistent with the measurement data.
□

Theoretically, Prop. 5.1 can be checked by applying the verification equation (4.20) based on the theorem of Prager-Oettli to each parameter vector $\Theta^* \in \mathbf{\Theta}^*$. However, to realize this approach in an algorithmic implementation it is necessary to draw n_{check} discrete samples from the continuous intervals. The number of discrete parameter vectors to check n_{check} thus becomes a relevant design parameter. According to Prop. 5.2, a single inconsistent vector falsifies the full consistency property. This leads to the requirement that n_{check} has to be very large to cover the given parameter range sufficiently. Even though a single evaluation of the theorem of Prager-Oettli is computationally very effective as stated in Chapter 4, the computation time rises proportionally with n_{check}. Additionally, as this thesis aims on calculating guaranteed results, the step size used to sample the interval type parameter vector needs to be very fine, even tending to zero. This high resolution needs to be applied to each component of the parameter vector. Afterwards it is used to build all possible combinations including the samples of the different components. Assuming a resolution of n_s samples on each component of $\mathbf{\Theta}^*$ leads to

$$n_{check} = n_s^{n_a^* + n_c^*} \tag{5.11}$$

applications of the theorem of Prager-Oettli . This number increases polynomial with n_s and leads to large computation times for sufficiently high resolutions. Thus the sampling based approach is computationally infeasible.

Restructuring the problem can be used to avoid the necessity to cover the whole parameter area. The verdicts can be calculated based on the vertexes $\mathcal{V}$ of the nominal parameter set only and then can be generalized to the whole nominal set if convexity properties are fulfilled.

The maximum number of points to check is thus reduced to

$$n_{check} = 2^{n_a^*+n_c^*} \tag{5.12}$$

which resembles a computationally feasible number, especially for low system orders n_a^* and n_c^*. The vertexes $\mathcal{V}$ of the hyperrectangle given by the interval type parameter vector $\boldsymbol{\Theta}^*$ can be determined according to Def. 5.4.

Definition 5.4 (Vertexes of a Hyperrectangle)
The nominal interval vector $\boldsymbol{\Theta}^* \in \mathbb{IR}^{n_a^*+n_c^* \times 1}$ *defines*

$$n = 2^{n_a^*+n_c^*} \tag{5.13}$$

vertexes $\mathcal{V} \in \mathbb{IR}^{n_a^*+n_c^* \times 1}$ *of a hyperrectangle that can be indexed using a decimal index* $v_{dec} \in \{0, 1, \ldots, n-1\}$*. The index is subsequently transformed to its binary representation* V_{bin} *that can be interpreted as* $(1 \times n_a^* + n_c^*)$ *dimensional vector were the* i*-th vector component is denoted as* $V_{bin}^{(i)}$*.*
The specific values of the vertexes $V_{v_{dec}}$ *can be generated by interpreting the binary index* V_{bin} *component wise for* $i \in \{1, 2, \ldots, n_a^* + n_c^*\}$ *and extracting the limits from the respective nominal parameter vector element:*

$$V_{v_{dec}}^{(i)} = \begin{cases} \underline{\Theta}^{*(i)} & , \mathit{if}\, V_{bin}^{(i)} = 0 \\ \overline{\Theta}^{*(i)} & , \mathit{if}\, V_{bin}^{(i)} = 1. \end{cases} \tag{5.14}$$

An illustration of Def. 5.4 is given in Example 5.1.

Example 5.1:
Consider the following (2×1) *interval vector with* $n_a^* = n_c^* = 1$

$$\boldsymbol{\Theta}^* = [[2,\ 3],\ [4,\ 6]]^T \tag{5.15}$$

The resulting rectangle has $n = 2^2 = 4$ *vertexes with* $v_{dec} \in \{0, 1, 2, 3\}$ *and* $i \in \{1, 2\}$ *leading to the indexes given in Tab. 5.1.*

Table 5.1: Vertexes of a hyperrectangle

Decimal index v_{dec}	*Binary index* V_{bin}	*Coordinates* $V_{v_{dec}}$ *according to (5.14)*
0	[0 0]	[2 4]
1	[0 1]	[2 6]
2	[1 0]	[3 4]
3	[1 1]	[3 6]

It is now possible to set up an alternative formulation of Prop. 5.1, that solves Problem 5.1 with vertex based full consistency.

Proposition 5.3 (Vertex Based Full Consistency)
The interval enclosure of the measurement data $[\mathbf{U}_{meas}, \mathbf{Y}_{meas}]$ *given for the discrete sampling points* $k = \{1, 2, \ldots, T\}$*, forming the regressor matrix* $\boldsymbol{A}_{meas}$ *and the measurement vector* $\boldsymbol{B}_{meas}$ *of a VO, is called to be full consistent with an interval type specification* S_i^**, if all vertexes* $\mathcal{V}$ *defined by the set of specified parameters* $\boldsymbol{\Theta}^*$ *are located in the same orthant and are part of the united solution set* $\sum_{\exists\exists}$ *given by the measurement data, i.e.*

$$(\mathcal{V} \subseteq \textstyle\sum_{\exists\exists}(\boldsymbol{A}_{meas}, \boldsymbol{B}_{meas})) \Leftrightarrow Full\ Consistency^{-} \Rightarrow Full\ Consistency. \quad (5.16)$$

Proof:
The united solution set can form various shapes, but it was shown by [Sha10] that the general AE-solution is convex within each orthant. As the united solution set is a special case of the general AE-solution, this property does also hold for $\sum_{\exists\exists}(\boldsymbol{A}_{meas}, \boldsymbol{B}_{meas})$.
The specified parameters $\boldsymbol{\Theta}^*$ and thus the resulting vertexes $\mathcal{V}$ are all located within the same orthant.
The theorem of Prager-Oettli can be checked for the $n = 2^{n_a^* + n_c^*}$ vertexes in finite time.
Based on the direct application of the definition of a convex set given in [Bro08, p. 662] follows:
If (3.65) holds for any two of the vertexes V_i and V_j, with $i, j \in \{1, 2, \ldots, n-1\}$, $i \neq j$, all vectors $\Theta = \lambda V_i + (1 - \lambda) V_j$, with $0 \leq \lambda \leq 1$ are also part of the united solution set. □

Example 5.2:

Assume the same setting as in Example 4.1. However the specification is now given as an interval type specification

$$S_i^* = \left\{ \boldsymbol{\Theta}^* = \left[[0.9725,\ 0.9925]\, ,\ [0.0665,\ 0.0685] \right]^T, n_a^* = 1, n_c^* = 1 \right\}. \tag{5.17}$$

The simulations are done using the linear discrete time ARX system (4.22) leading to the same measurement data as given in Fig. 4.2 in Example 4.1. The resulting verification setting is depicted in Fig. 5.1. The nominal parameters given in S_i^* are depicted as green square. It can be seen that all four vertexes $\mathcal{V} = \{V_0, V_1, V_2, V_3\}$ are located within the united solution set given by the measurement data. The specification and the measurement are thus guaranteed to be full consistent according to Prop. 5.3.

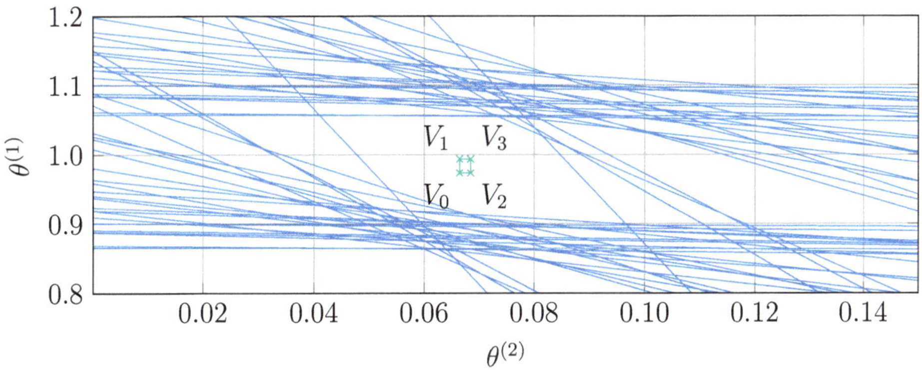

Figure 5.1: Example setting for an interval type specification S_i^*

5.2 Interval Type Basic Consistency

Until now the set of VO behavior was assumed to be considerably larger than the specification. When using interval type specifications this is not necessarily the case. It is possible that the specification set is of the same size as the set of VO behavior or even larger. Therefore it is not longer possible to enclose the whole specification in the VO behavior. The resulting verification question can be formulated as follows:

Problem 5.2 (Interval Type Basic Consistency)
Is the nominal system, specified by an interval type specification

$$S_i^* = \{\mathbf{\Theta}^*, n_a^*, n_c^*, U_{init}^*, Y_{init}^*\}, \tag{5.18}$$

basic consistent with the input-output behavior given by the interval type enclosures of T *measurement values*

$$[\mathbf{U}_{meas}, \mathbf{Y}_{meas}] = \left[\langle \mathbf{u}_{meas,k}\rangle_{k=1}^{T}, \langle \mathbf{y}_{meas,k}\rangle_{k=1}^{T}\right] \tag{5.19}$$

i.e. is there at least one parameter vector $\Theta^* \in \mathbf{\Theta}^*$ *that fulfills Prop. 4.1?*

The Venn chart of a basic consistent setting is depicted in Fig. 5.2. Due to the inner enclosure of the VO behavior the achieved verdict is still type II error free. The consistent set is given by the green shaded square.
This set it formally stated in Prop. 5.4 and solves Problem 5.2.

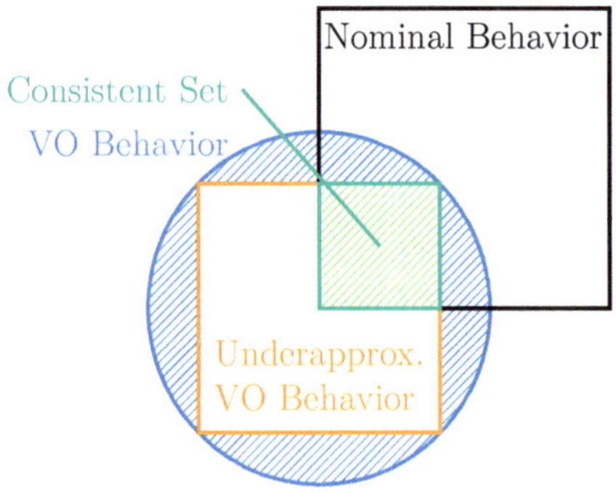

Figure 5.2: Basic consistent result for a large specification

Proposition 5.4 (Interval Type Basic Consistency)
The interval enclosure of the measurement data $[\mathbf{U}_{meas}, \mathbf{Y}_{meas}]$ *given for the discrete sampling points* $k \in \{1, 2, \ldots, T\}$, *forming the regressor matrix* $\boldsymbol{A}_{meas}$ *and the measurement vector* $\boldsymbol{B}_{meas}$ *of a VO, is called basic consistent with an interval type specification* S_i^*, *if there is a nonempty consistent set, i.e. a nonempty intersection between the nominal set* $\boldsymbol{\Theta}^*$ *and the united solution set* $\sum_{\exists\exists}(\boldsymbol{A}_{meas}, \boldsymbol{B}_{meas})$

$$(\boldsymbol{\Theta}^* \cap \textstyle\sum_{\exists\exists}(\boldsymbol{A}_{meas}, \boldsymbol{B}_{meas}) \neq \emptyset) \Leftrightarrow Basic\,Consistency^- \Rightarrow Basic\,Consistency. \tag{5.20}$$

Proof:
All parameter vectors included in the interval specification $\boldsymbol{\Theta}^*$ are suitable representations of the correct system behavior. Thus S_i^* can be interpreted as a set of direct specifications S_d^*. Each of these direct specifications $S_d^* \in S_i^*$ can be used to check consistency according to Prop. 4.1. If there is at least one full consistent direct specification included in the interval specification, the VO behavior can be explained by this parameter and the VO is denoted as basic consistent. □

$Basic\,Consistency^-$ is sufficient for the genuine system, as that there is at least one parameter within the nominal set that is able to explain the measurement data.

5.2.1 Algorithmic Solutions

A straight forward approach to check basic consistency uses the vertexes only. However, the shape of the resulting consistent set will change if one or more vertexes are inconsistent. This change is depicted exemplary for the $2D$ case in Fig. 5.3.

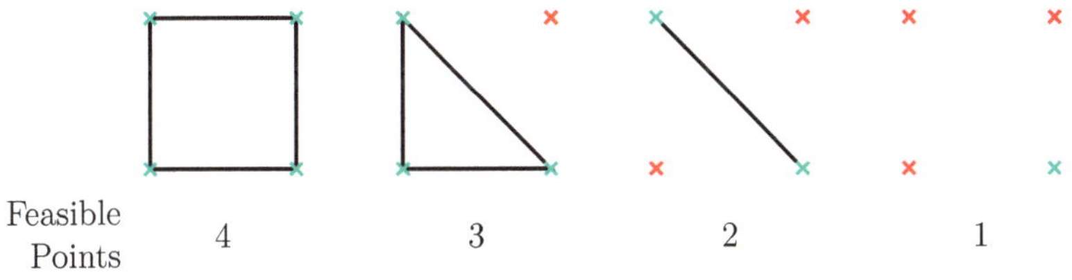

Figure 5.3: Degradation of the consistent set for different consistent vertexes (green)

The shape changes from a rectangle, in case all four vertexes are part of the consistent set, to a single point if only one vertex shows consistency. The main drawback of this procedure is that there might be a consistent set, even if no initial vertex is consistent. This situation is exemplary depicted in Fig. 5.5.

Also, the basic shape of the resulting consistent set changes which can be a disadvantage for the further algorithmic processing of the result.
A solution for the first problem is given by checking more points that are not a vertex. However this leads to the sampling based approach as introduced in the previous section, with the respective runtime limitations explained there.
The problem of finding the right resolution in the sampling based approach can be solved using optimization methods. The idea is to use the logic of optimization algorithms to guide the sampling process.[6] Optimization procedures can be used to determine which discrete points to check if one or more vertexes of the specification are not element of the consistent set.
To use optimization methods for the choice of sampling points the problem can be reformulated to a feasibility problem. The interval type measurement data is used to set up the constraints that frame the united solution set $\sum_{\exists\exists}$. As derived in the previous chapter, every parameter within the united solution set $\sum_{\exists\exists}$ is able to explain the measurement data. The constraints limit the search area of the feasibility problem. Feasible solutions need to fulfill all constraints. Parameter values that are located outside the united solution set $\sum_{\exists\exists}$ are not feasible with respect to the constraints.

The interval type specification S_i^* includes the nominal parameter vector $\mathbf{\Theta}^*$ which can also be denoted as nominal set $\mathcal{N}$. The nominal set represents the maximum area of potentially consistent parameters. This initial restrictions can be stated in terms of linear inequality constraints as follows:

Definition 5.5 (Constraints Given by the Nominal Set)
The interval type parameter vector $\mathbf{\Theta}^*$ given in the specification S_i^* can be used to set up the set $c_{\mathcal{N}}$ of $2(n_a^* + n_c^*)$ linear inequality constraints that restrict the feasibility problem to the nominal set $\mathcal{N}$:

$$c_{\mathcal{N}}^{(i)}(\mathbf{\Theta}) := \underline{\theta}^{*(i)} - \underline{\theta}^{(i)} \leq 0 \tag{5.21}$$

$$c_{\mathcal{N}}^{(n_a^*+n_c^*+i)}(\mathbf{\Theta}) := -\overline{\theta}^{*(i)} + \overline{\theta}^{(i)} \leq 0 \tag{5.22}$$

with the number of parameters $i \in \{1, 2, \ldots, n_a^* + n_c^*\}$.

[6] Suitable optimization methods are e.g. grid search, golden section search or dichotomous search [Wil64] if there is only minimal information available. If there is further knowledge about the shape of the problem, there are more sophisticated algorithms that can direct the search effort very efficiently into the relevant regions, e.g. Newton method, simplex method or interior point method [Noc06].

The united solution set $\sum_{\exists\exists}$ can be reformulated in terms of the interval type measurement data as linear matrix inequality constraints (LMI) as given in Def. 5.6.

Definition 5.6 (Constraints Given by the Measurement Data)
The measurement data $[\mathbf{U}_{meas}, \mathbf{Y}_{meas}] = \left[\langle \mathbf{u}_{meas,k}\rangle_{k=1}^{T}, \langle \mathbf{y}_{meas,k}\rangle_{k=1}^{T}\right]$ *forming the regressor matrix* $\boldsymbol{A}_{meas}$ *and the measurement vector* $\boldsymbol{B}_{meas}$ *of a VO, can be used to set up the set* $c_{\mathcal{M}}$ *of* $2\,(T - \max{(n_a^*, n_c^*)})$ *linear inequalities*

$$c_{\mathcal{M}}^{(i)}(\boldsymbol{\Theta}) := -\underline{(\boldsymbol{A}_{meas} \cdot \boldsymbol{\Theta})}^{(i)} + \underline{B}_{meas}^{(i)} \leq 0 \tag{5.23}$$

$$c_{\mathcal{M}}^{(T-\max{(n_a^*,n_c^*)}+i)}(\boldsymbol{\Theta}) := \overline{(\boldsymbol{A}_{meas} \cdot \boldsymbol{\Theta})}^{(i)} - \overline{B}_{meas}^{(i)} \leq 0 \tag{5.24}$$

with the number of rows in the regressor matrix $i \in \{1, 2, \ldots, T - \max{(n_a^*, n_c^*)}\}$*, i.e. the number of system equations instantiated by different measurement points.*

Each pair of constraints represents the upper and lower bound of the solution set derived from one line of the ILES (3.46). They can be interpreted as hyperstripes in the parameter space, leading to the setting shown in Fig. 5.4 for the $2D$ case. Using constraints according to Def. 5.6 limits the search area of the optimization algorithm to the inner approximation and thus guarantees that there are no type II errors possible in the resulting consistent set. The consistent set is given by the shaded region.

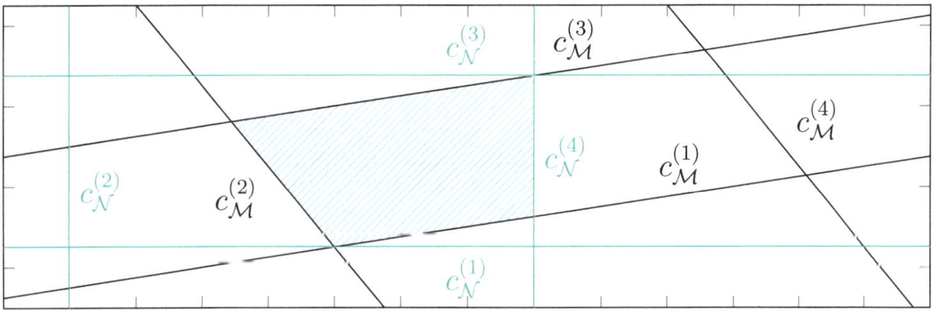

Figure 5.4: Exemplary constraints of the feasibility problem in $2D$

Note that the number of constraints in $c_{\mathcal{M}}$ is directly related with the number of used sampling points T and that the number of parameters $n = n_a^* + n_c^*$ is of minor importance. However there have to be at least $T = k_{min} = \max(n_a^*, n_c^*) + 1$ samples in the measurement to set up the first hyperstripe. It is now possible to determine an alternative solution for Problem 5.2, based on the constraints of Def. 5.5 and Def. 5.6.

Proposition 5.5 (Feasibility Based Basic Consistency)
The interval enclosure of the measurement data $[\mathbf{U}_{meas}, \mathbf{Y}_{meas}]$ *given for the discrete sampling points* $k \in \{1, 2, \dots, T\}$*, forming the regressor matrix* $\boldsymbol{A}_{meas}$ *and the measurement vector* $\boldsymbol{B}_{meas}$ *of a VO, is called basic consistent with an interval type specification* S_i^**, if the consistent set* $\mathcal{C}$ *is nonempty, i.e. if there is at least one solution* $\tilde{\Theta}$ *that fulfills all constraints*

$$\left(\left(c_{\mathcal{N}}(\tilde{\Theta}) \leq 0\right) \wedge \left(c_{\mathcal{M}}(\tilde{\Theta}) \leq 0\right)\right) \Leftrightarrow Basic\, Consistency^{-} \Rightarrow Basic\, Consistency \tag{5.25}$$

Proof:
According to Prop. 5.4 there needs to be at least one parameter that is part of the united solution set $\sum_{\exists\exists}$ as well as part of the parameter set given in the interval type specification S_i^* to ensure basic consistency. All parameters that fulfill the constraint set $c_{\mathcal{M}}$ of Def. 5.6 are part of the united solution set $\sum_{\exists\exists}$. All parameters that fulfill the constraint set $c_{\mathcal{N}}$ of Def. 5.5 are part of the parameter set given in the interval type specification S_i^*. If there is at least one parameter vector $\tilde{\Theta}$ that fulfills $c_{\mathcal{N}}$ and $c_{\mathcal{M}}$, all conditions for basic consistency are fulfilled. □

The resulting situation is exemplary depicted in Fig. 5.5. It can be seen that no initial vertex shows consistency, as none of them is part of the underapproximation. Therefore there is no vertex based basic consistency. Feasibility based basic consistency can be achieved as there is a consistent set within the inner approximation and the nominal set. The consistent set is depicted as the shaded green area.

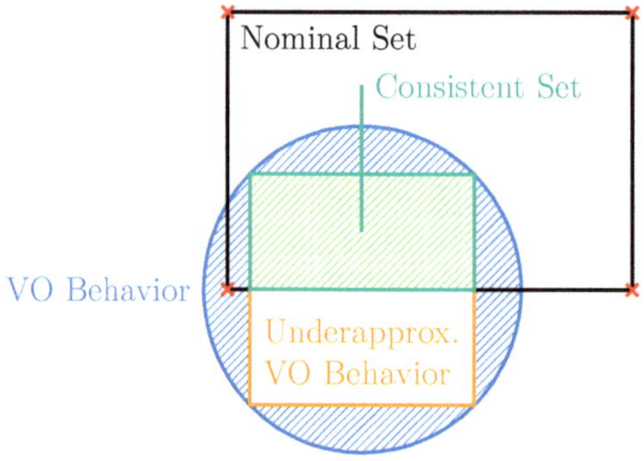

Figure 5.5: Example setting showing feasibility based consistency

5.3 Conclusion

This chapter introduced the extension of the Kaucher based method to interval type specifications. The resulting nominal set is required to be located within one orthant. The definition of full consistency is still applicable if all parameters given in the interval type specification are part of the united solution set. A straight forward approach was introduced that used the orthant wise convexity of the united solution set to define full consistency based only on the vertexes of the specified parameter set.
Nevertheless, full consistency is a rather strict criterion and it is likely that there are some specified parameters that are part of the united solution and some that are not. Following the concept of basic consistency, it is sufficient to show that there is at least one parameter vector that is part of the united solution set as well as of the specified parameter range. This property can be verified by checking individual arbitrary points for consistency. The choice of these points can be structured by using the concept of a feasibility problem. Therefore the notions defining the united solution as well as the specified parameter set are transferred to linear matrix inequality constraints. If there is a solution of the feasibility problem, the VO and the specification are guaranteed to be basic consistent.
The method is still free of type II errors as the feasibility based consistent set is constrained by the genuine united solution set.

6 Guaranteed Verification of Hybrid Systems

Switched hybrid systems consist of two distinct parts with different properties and modeling goals. The dynamic part is used to model the plant dynamics as introduced and used in the previous chapters. The additional discrete event part models the superimposed switching logic. Based on logical rules, the discrete event part can activate different discrete states that will show different dynamic behavior. Different operation modes can thus be modeled as several subsystems, showing individual behavior. All subsystems are interconnected by the discrete event switching mechanism.
The switched hybrid system structure is depicted in Fig. 6.1 and is formally introduced in this section[7].

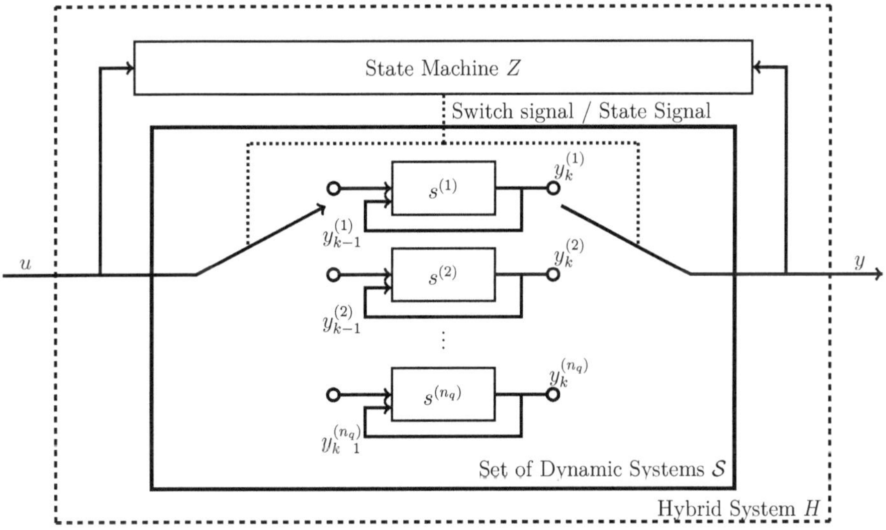

Figure 6.1: Structure of the hybrid system model H

The hybrid system $\mathcal{H}$ consists of a set of dynamic subsystems $\mathcal{S}$ and a superimposed switching mechanism represented by the state machine Z.

The subsystems $s^{(i)} \in \mathcal{S}$ show different behavior based on an individual parametrization. The state machine produces a switch signal which resembles its current discrete state. This signal is used to control an input and an output switch that determines which subsystem is activated.

7 To improve readability the term "hybrid system" is used instead of "switched hybrid system" throughout this thesis.

The activated subsystem is fed with the general input signal and the resulting output signal is connected to the output of the hybrid system. It is possible to use the optional input and output values that are specified in the nominal system to start the active subsystem after a switch. Otherwise the current input and output values are kept across the switch. The input and output signals are also fed to the state machine where they are used to update the discrete state of the discrete event system. Due to this extended structure there are additional subjects included in the verification question that will be covered in this chapter. First the formalized model structure needs to be appended to a hybrid formulation. Therefore the discrete event part is modeled in the following, based on [Cas99, p. 66ff].
Then the verification problem is split in two subproblems: verification of the dynamic part and verification of the discrete event part. If both parts are verified individually, the next step is to examine their connection and interaction. This is done in three steps of increasing complexity.
Initially, the setting is simplified by assuming the discrete state to be measurable. This setting is used to introduce the basic hybrid method. Second this assumption is dropped such that the current active discrete state needs to be determined for a given set of switches. Finally an algorithm is developed that is able to determine the switching times and the active discrete states from measured input and output data only.

The necessary knowledge for the verification procedure is thus the same as in the previous chapters except that there are now several nominal systems and an additional specification of the discrete event system part.

Definition 6.1 (Discrete State)
A discrete state

$$q^{(i)} \in \mathcal{Q} := \left\{ q^{(1)}, q^{(2)}, \dots, q^{(n_q)} \right\} \tag{6.1}$$

is a vertex of a graph e.g. of a state machine. It is part of a given set of discrete states $\mathcal{Q}$.

Note that for the ease of notation the “discrete state” is called “state” in the reminder of this thesis.

To implement logical conditions and constraints in the state machine a set of events is defined:

Definition 6.2 (Event)
There is a set of events

$$\mathcal{E} := \left\{ e^{(1)}, e^{(2)}, \ldots, e^{(n_e)} \right\}. \tag{6.2}$$

Each event $e^{(i)}$ *is dependent on specific activation limits*

$$\boldsymbol{l}^{(i)} = \left[\underline{l}^{(i)}, \, \overline{l}^{(i)} \right], \tag{6.3}$$

with $i \in \{1, 2, \ldots, n_e\}$*, defined for a given enabler signal* $W := \langle w_k \rangle_{k=1}^{T}$*. The event* $e^{(i)}$ *is defined to be active as long as the value of the enabler signal* W *lies within the given thresholds*

$$w_k \in \boldsymbol{l}^{(i)}. \tag{6.4}$$

Note that Def. 6.2 does not pose any conditions on the activation limits. Therefore it is possible that several events are active at the same time. In the context of this thesis, events are allowed to be active for several time steps, e.g. as long as the enabler signal stays within the specified limits.
The transitions of the state machine are defined by a transition function.

Definition 6.3 (Transition Function)
A transition function

$$f : \mathcal{Q} \times \mathcal{E} \rightarrow \mathcal{Q} \tag{6.5}$$

represents a directed connection between two states, labeled by an event. In general f *is a partial function on its domain.*

The notation $t^{(1)} : f\left(q^{(1)}, e^{(1)}\right) = q^{(2)}$ means that transition $t^{(1)}$ forms a directed connection from state $q^{(1)}$ to state $q^{(2)}$, dependent on event $e^{(1)}$.
A transition can change the state of a state machine if the assigned event is active, but not necessarily has to. This is due to the fact that several events can be active at any given time, but there is not more than one transition allowed to conduct a switch. However it is also possible that the state does not change even though there are several activated transitions.

Based on the above definitions it is possible to set up the state machine representing the discrete event part.

Definition 6.4 (State Machine)
A state machine is defined to be given by the 4-Tupel

$$Z := \left\{ \mathcal{Q}, \mathcal{E}, f, q^{(1)} \right\}, \tag{6.6}$$

with a finite set of states $\mathcal{Q}$*, a finite set of events* $\mathcal{E}$*, a transition function* f *and an initial state* $q^{(1)}$*.*

The state of the discrete event part can be used to determine the system orders and parameters of the dynamic part necessary to set up a system according to Def. 5.2. This leads to the definition of a state dependent, discrete time, linear, interval type system:

Definition 6.5 (State Dependent Discrete Time Linear Interval Type System)
The state dependent, discrete time, linear, interval type system can be modeled as

$$s^{(q_k)} := y_k = -\sum_{i=1}^{n_a(q_k)} \boldsymbol{a}_i(q_k) y_{k-i} + \sum_{i=1}^{n_c(q_k)} \boldsymbol{c}_i(q_k) u_{k-i} \tag{6.7}$$

with the discrete time input u_k*, output* y_k *and state* q_k*. The input and output orders* $n_a(q_k)$ *and* $n_c(q_k)$ *as well as the interval type system parameters* $\boldsymbol{\Theta} = \left[\boldsymbol{a}_1(q_k), \boldsymbol{a}_2(q_k), \dots, \boldsymbol{a}_{n_a(q_k)}(q_k), \boldsymbol{c}_1(q_k), \boldsymbol{c}_2(q_k), \dots, \boldsymbol{c}_{n_c(q_k)}(q_k)\right]$ *are dependent on the current state* q_k*. All subsystems* $s^{(q_k)}$ *form the set of subsystems*

$$\mathcal{S} = \left\{ s^{(1)}, s^{(2)}, \dots, s^{(n_q)} \right\}. \tag{6.8}$$

Each dynamic subsystem is directly linked with a discrete state. Therefore the state dependent dynamic subsystem $s^{\left(q^{(i)}\right)}$ is denoted by $s^{(i)}$ for the ease of notation. It is now possible to define the overall hybrid system.

Definition 6.6 (Hybrid System)
A hybrid system $\mathcal{H}$ consists of two system parts:

Discrete Event Part *The superimposed switching mechanism given by a state machine*

$$Z = \left\{\mathcal{Q}, \mathcal{E}, f, q^{(1)}\right\}, \tag{6.9}$$

according to Def. 6.4.

Dynamic Part *The discrete time linear interval type systems are given by a finite set of subsystems*

$$\mathcal{S} = \left\{s^{(1)}, s^{(2)}, \ldots, s^{(n_q)}\right\} \tag{6.10}$$

where each subsystem $s^{(q_k)}$ is active if and only if q_k is the current state. The subsystems $s^{(q_k)}$ are defined according to Def. 6.5.

In general it is not possible to measure the current state of the state machine. However, for didactic reasons Assumption 6.1 is introduced and later dropped.

Assumption 6.1 (Measured State Signal)
The current state of the state machine can be measured and it is correctly given in the state signal

$$Q_{meas} := \langle q_{meas,k} \rangle_{k=1}^{T}. \tag{6.11}$$

The measured state signal Q_{meas} consists of several segments with different active states. A change in the active state is called switch.

Definition 6.7 (Switch)
The first time index k at which a new state is active i.e.

$$q_{meas,k} \neq q_{meas,k-1}, \tag{6.12}$$

is called switch k_τ.

The switches within a state signal are additionally indexed in chronological order $k_{\tau,i}$, $i \in \{1, 2, \ldots, n_{switch}\}$ with the total number of switches denoted by n_{switch}. The time index given by the first sampling point of the measurement data represents the first occurrence of the initial state and is thus defined to be $k_{\tau,1} = 1$.

The time index right before a switch, i.e. the last time index the current state is active, is called end of the current segment $k_{\mathcal{T}',i} = k_{\mathcal{T},i+1} - 1$. The end of the last segment is given by the last time index of the measurement data T which leads to $k_{\mathcal{T}',n_{switch}} = T$. A schematic sketch for $n_{switch} = 3$ is given in Fig. 6.2.

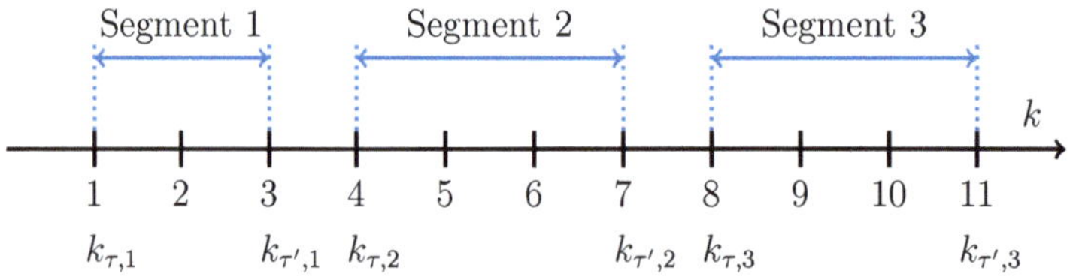

Figure 6.2: Schematic view of switches

6.1 Verification of Hybrid Systems with Mapped State Signal

The introduced definitions can be used to set up the specification of a hybrid system.

Definition 6.8 (Specification of an Interval Type Hybrid System)
An interval type specification of a hybrid system according to Def. 6.6 is given by

$$S^*_{H,i} := \{Z^*, S^*_i\} \tag{6.13}$$

with the nominal state machine Z^* according to Def. 6.4 and the set of nominal dynamic systems S^*_i, according to Def. 5.1.

A special case of Def. 6.8 is given if the parameters of the dynamic subsystems are point real values.

Definition 6.9 (Specification of a Point Real Hybrid System)
A point real specification of a hybrid system according to Def. 6.6 is given by

$$S^*_{H,d} := \{Z^*, S^*_d\} \tag{6.14}$$

with the nominal state machine Z^* according to Def. 6.4 and the set of nominal dynamic systems S^*_d, according to Def. 4.2.

The dynamic part of a system according to Def. 6.9 is also called linear time variant system with piecewise constant parameters.

The verification method developed in this chapter is again based on interval enclosures of the input and output measurement data

$$[\boldsymbol{U}_{meas}, \boldsymbol{Y}_{meas}] = \left[\langle \boldsymbol{u}_{meas,k}\rangle_{k=1}^{T}, \langle \boldsymbol{y}_{meas,k}\rangle_{k=1}^{T}\right] \tag{6.15}$$

extended by a measured state signal Q_{meas} according to Assumption 6.1. The state signal Q_{meas} leads to a set of measured states $\mathcal{Q}_{meas}$. It is assumed that the elements of the set of measured states can be mapped on the set of specified states. In this case, the set is called mapped set of states according to Def. 6.10.

Definition 6.10 (Mapped Set of States)
The set of measured states $\mathcal{Q}_{meas}$, consisting of the unique values of the state signal $Q_{meas} = \langle q_{meas,k}\rangle_{k=1}^{T}$, is called mapped with the specification $S_{H,i}^{*}$, if all measured states $q_{meas}^{(j)} \in \mathcal{Q}_{meas}$ with $j \in \{1, 2, \ldots, |\mathcal{Q}_{meas}|\}$ can be mapped to an equivalent nominal state $q^{(i(j))*} \in \mathcal{Q}^{*}$ given in the specification i.e.

$$q_{meas}^{(j)} = q^{(i(j))*}, \; j \in \{1, 2, \ldots, |\mathcal{Q}_{meas}|\}. \tag{6.16}$$

As $S_{H,d}^{*}$ is a special case of $S_{H,i}^{*}$, Def. 6.10 is also valid for $S_{H,d}^{*}$. In case $\mathcal{Q}_{meas}$ is a mapped set of states, the state signal Q_{meas} is called mapped state signal. If the mapped state $q^{i(j)*}$ is known, the respective mapped nominal subsystem $s^{i(j)*}$ is also known. The hybrid verification problem can now be formulated as follows:

Problem 6.1 (Mapped Set of States Based Point Real Hybrid Consistency)
Is the nominal hybrid system, specified by a point real hybrid specification

$$S_{H,d}^{*} = \{Z^{*}, S_{d}^{*}\} \tag{6.17}$$

consistent with the input-output behavior given by the interval type enclosures of T measurement values

$$[\boldsymbol{U}_{meas}, \boldsymbol{Y}_{meas}] = \left[\langle \boldsymbol{u}_{meas,k}\rangle_{k=1}^{T}, \langle \boldsymbol{y}_{meas,k}\rangle_{k=1}^{T}\right] \tag{6.18}$$

and the measured mapped state signal Q_{meas}, i.e. can the measurement data be explained by the nominal system?

The problem can be solved by tackling the two system parts individually. This is possible due to an implicit connection given by the matching of the discrete states which is done based on the dynamic parameters. First the dynamic subsystem is considered and verified in the next section. Afterwards the verification of the discrete event part is introduced. Finally the results are combined to verify the overall hybrid system.

6.1.1 Verification of the Dynamic Subsystems

To verify the individual dynamic subsystems it is necessary to split the interval type measurement data $[\mathbf{U}_{meas}, \mathbf{Y}_{meas}]$ based on the information given in the mapped state signal. The resulting segments

$$\left\langle \left[\mathbf{U}_{meas}^{(j)}, \mathbf{Y}_{meas}^{(j)}\right] \right\rangle_{j=1}^{n_{switch}} = \left\langle \left[\langle \mathbf{u}_{meas,k} \rangle_{k=k_{\tau,j}}^{k_{\tau',j}}, \langle \mathbf{y}_{meas,k} \rangle_{k=k_{\tau,j}}^{k_{\tau',j}}\right] \right\rangle_{j=1}^{n_{switch}} \tag{6.19}$$

can be verified individually against the respective specification in the set of subsystems $\mathcal{S}_d^*$. It is possible to verify the individual dynamic behavior[8] of each state present in the measurement data $q_{meas}^{(j)} \in Q_{meas}$ as defined in Prop. 6.1.

Proposition 6.1 (Dynamic Consistency of a Segment)
The interval type enclosure of the measurement data is split into $j \in \{1, 2, \ldots, n_{switch}\}$ parts, given by the segments $\left[\mathbf{U}_{meas}^{(j)}, \mathbf{Y}_{meas}^{(j)}\right]$. Each segment represents a specific state $q_{meas}^{(j)}$, a regressor matrix $\boldsymbol{A}_{meas}^{(j)}$ and a measurement vector $\boldsymbol{B}_{meas}^{(j)}$. The segment j is dynamic consistent with the respective mapped subsystem $s^{(i(j))*} \in \mathcal{S}_d^*$, if the specified system parameters $\Theta^{(i(j))*} = \left[a_1^*(i(j)), a_2^*(i(j)), \ldots, a_{n_a^*(i(j))}^*(i(j)), c_1^*(i(j)), c_2^*(i(j)), \ldots, c_{n_c^*(i(j))}^*(i(j))\right]^T$ are part of the united solution set $\sum_{\exists\exists}^{(j)} = \sum_{\exists\exists}(\boldsymbol{A}_{meas}^{(j)}, \boldsymbol{B}_{meas}^{(j)})$, i.e. if

$$\Theta^{(i(j))*} \in \sum\nolimits_{\exists\exists}^{(j)} . \tag{6.20}$$

Proof:
The measured state signal Q_{meas} includes the true active states. It is thus guaranteed that only measurement data generated by subsystem $s_{meas}^{(j)}$ based on the active state $q_{meas}^{(j)}$ is included in $\left[\mathbf{U}_{meas}^{(j)}, \mathbf{Y}_{meas}^{(j)}\right]$ and that this data is not corrupted by measurement generated by other subsystems. The current state $q_{meas}^{(j)}$ is a mapped state according to Def. 6.10 and thus the connection between measurement and specification is also correct and the respective nominal subsystem $s^{(i(j))*}$ is known.
Using this information, the setting can be reduced to the time invariant consistency problem given in Problem 4.1 for segment j and subsystem $s^{(i(j))*}$ and time invariant consistency can be checked according to Prop. 4.1 which proves Prop. 6.1. □

The considerations are now extended to the complete set of measurement data, consisting of several segments.

[8] As a direct specification $\mathcal{S}_d^*$ is used in the definition, "dynamic consistency" means "time invariant full consistency" throughout this chapter.

Proposition 6.2 (Dynamic Consistency of the Measurement Data)
The mapped state signal Q_{meas} *and the segmented measurement data* $\left\langle\left[\mathbf{U}_{meas}^{(j)}, \mathbf{Y}_{meas}^{(j)}\right]\right\rangle_{j=1}^{n_{switch}}$ *of a VO, leading to the state* $q_{meas}^{(j)}$, *the regressor matrices* $\boldsymbol{A}_{meas}^{(j)}$ *and the measurement vectors* $\boldsymbol{B}_{meas}^{(j)}$ *with* $j = \{1, 2, \ldots, n_{switch}\}$, *are dynamic consistent with a set of direct specifications* S_d^*, *if there is dynamic consistency of each segment given in the measurement data with its respective mapped subsystem* $s^{(i(j))*}$ *i.e.*

$$\Theta^{(i(j))*} \in \Sigma_{\exists\exists}^{(j)}, \ \forall j \in \{1, 2, \ldots, n_{switch}\}. \tag{6.21}$$

Proof:
Prop. 6.2 results straight forward by applying Prop. 6.1 to all $j = \{1, 2, \ldots, n_{switch}\}$ segments given in the segmented measurement data $\left\langle\left[\mathbf{U}_{meas}^{(j)}, \mathbf{Y}_{meas}^{(j)}\right]\right\rangle_{j=1}^{n_{switch}}$. □

It is possible to reformulate this proposition to get the inverse relation similar to Prop. 5.2.

Proposition 6.3 (Inverse of Dynamic Consistency of the Measurement Data)
The mapped state signal Q_{meas} *and the segmented measurement data* $\left\langle\left[\mathbf{U}_{meas}^{(j)}, \mathbf{Y}_{meas}^{(j)}\right]\right\rangle_{j=1}^{n_{switch}}$ *of a VO, leading to the state* $q_{meas}^{(j)}$, *the regressor matrices* $\boldsymbol{A}_{meas}^{(j)}$ *and the measurement vectors* $\boldsymbol{B}_{meas}^{(j)}$ *with* $j = \{1, 2, \ldots, n_{switch}\}$ *are called dynamic inconsistent with a set of direct specifications* S_d^*, *if there is at least one segment* $j \in \{1, 2, \ldots, n_{switch}\}$ *given in the measurement data that does not show dynamic consistency with its respective mapped subsystems* $s^{(i(j))*}$ *according to Prop. 6.1, i.e.*

$$\exists j = \{1, 2, \ldots, n_{switch}\} \ \middle| \ \Theta^{(i(j))*} \notin \Sigma_{\exists\exists}^{(j)}. \tag{6.22}$$

Proof:
For dynamic consistency of the measurement data according to Prop. 6.2 it is necessary that all segments of the measurement data are dynamic consistent according to Prop. 6.1. If there is a segment that does not show dynamic consistency, this segment can not be explained by the specification. Thus there is unspecified behavior and it is not possible to explain the whole measurement data by the specification. □

Note that these propositions are based on the segments given in the measurement data. However it is possible that there is nominal behavior that is not present in the measurement data. Though this will not change the dynamic consistency result for the measurement, it will influence the discrete event verification result introduced in the next section.

Also note that a mapped state signal is used in Prop. 6.2 and Prop. 6.3. This means that all segments given in the measurement data can be mapped to the specification, as stated in Def. 6.10.

6.1.2 Verification of the Discrete Event System

The next step in the verification of the hybrid system is given by regarding the discrete event part. Therefore it is necessary to determine the state machine Z_{meas} that generated the measurement data. The active states given in the mapped state signal Q_{meas} represent a trace of the unknown generating state machine Z_{meas}. The system dynamics represent the state of the system and not an emission of a state or an event. Therefore it is in this case possible to reconstruct Z_{meas} based on this trace. The reconstructed generating state machine Z_{meas} can then be compared with the specified state machine Z^* by comparing the corresponding states and transitions.

Proposition 6.4 (Full State Consistency)
The mapped state signal Q_{meas} given for the discrete sampling points $k = \{1, 2, \dots, T\}$, leading to the mapped set of states $\mathcal{Q}_{meas}$ of the discrete event part Z_{meas} of a VO is called full state consistent with the nominal state machine Z^*, if

$$\mathcal{Q}^* = \mathcal{Q}_{meas}. \tag{6.23}$$

Proof:
All nominal states $\mathcal{Q}^*$ are given in the specification of the state machine Z^*. The states implemented in Z_{meas} are given in the mapped set of states $\mathcal{Q}_{meas}$. According to Def. 6.10, this means that both sets are defined on the same elements. Full state consistency means that exactly the specified states are given in the measurement data, i.e. that both sets contain the same elements. This comparison is given in (6.23) and proves full state consistency. □

Due to a measurement scenario that does not cover all states it is possible that not all dynamics are present in the measurement data. This case leads to partial state consistency.

Proposition 6.5 (Partial State Consistency)
The mapped state signal Q_{meas} given for the discrete sampling points $k = \{1, 2, \dots, T\}$, leading to the mapped set of states $\mathcal{Q}_{meas}$, of the discrete event part Z_{meas} of a VO is called partial state consistent with the nominal state machine Z^*, if

$$\mathcal{Q}^* \supset \mathcal{Q}_{meas}. \tag{6.24}$$

Proof:
Based on Prop. 6.4 and Def. 6.10, $\mathcal{Q}_{meas}$ and $\mathcal{Q}^*$ are defined on the same elements, i.e. there is a mapping $q_{meas}^{(j)} = q^{(i(j))*}$ for $j \in \{1, 2, \ldots, |\mathcal{Q}_{meas}|\}$. If additionally (6.24) holds,

$$q_{meas}^{(j)} \in \mathcal{Q}_{meas} \Rightarrow q_{meas}^{(j)} = q^{(i(j))*} \in \mathcal{Q}^* \tag{6.25}$$

holds as well. This means that all implemented states $q_{meas}^{(j)} \in \mathcal{Q}_{meas}$ are also part of the specification $\mathcal{Q}^*$. Thus there are only specified states in the measurement data. However, there are states in the specification that are not part of the implementation and prevent full state consistency. The discrete event part Z_{meas} of a VO is hence called partial state consistent. □

Partial consistency of the discrete event part Z_{meas} of a VO is a hint to improve the measurement scenario or to collect more measurement data.
If there is neither full nor partial consistency, the system is state inconsistent according to Prop. 6.6.

Proposition 6.6 (State Inconsistency)
The state signal Q_{meas} given for the discrete sampling points $k = \{1, 2, \ldots, T\}$, leading to the set of states $\mathcal{Q}_{meas}$, of the discrete event part Z_{meas} of a VO is called state inconsistent with the nominal state machine Z^*, if

$$\mathcal{Q}^* \not\supseteq \mathcal{Q}_{meas}. \tag{6.26}$$

Proof:
If (6.26) holds, there are implemented states $q_{meas}^{(j)} \in \mathcal{Q}_{meas}$ that are not part of the specification $\mathcal{Q}^*$. Thus there are unspecified states in the measurement data. The discrete event part Z_{meas} of a VO is hence called state inconsistent. □

Prop. 6.6 can be connected with the inverse of dynamic consistency of the measurement data as given in Prop. 6.3. If there is an additional state $q_{meas}^{(j)} \notin \mathcal{Q}^*$, the respective measurement data $\left[\mathbf{U}_{meas}^{(j)}, \mathbf{Y}_{meas}^{(j)}\right]$ cannot be explained by any $s^{(i)*}$ within the specification. In this case there is both, dynamic inconsistency according to Prop. 6.3 and state inconsistency according to Prop. 6.6.

The second part of the discrete event system to be verified is the transition function. Additionally to the mapped state signal[9] $\langle q_{meas,k} \rangle_{k=1}^{T-1}$, the set of measured events $\mathcal{E}_{meas}$ needs to be obtained. According to Def. 6.2 there is an active event $e_{meas,k}^{(i)} \in \mathcal{E}_{meas}$ if $w_{meas,k} \in \boldsymbol{l}^{(i)*}$.

9 The last measurement value for $k = T$ can not be evaluated as there is no following state $k = T + 1$.

Full transition consistency is then defined as follows:

Proposition 6.7 (Full Transition Consistency)
The mapped state signal $\langle q_{meas,k} \rangle_{k=1}^{T-1}$ *and the set of events* $\mathcal{E}_{meas}$ *of the discrete event part* Z_{meas} *of a VO are called full transition consistent with the nominal state machine* Z^*, *if*

$$\left(\forall q_{meas,k} \in \langle q_{meas,k} \rangle_{k=1}^{T-1}\right) \wedge \left(\exists e_{meas,k}^{(i)} \in \mathcal{E}_{meas}\right) : \left(f^*(q_{meas,k}, e_{meas,k}^{(i)}) = q_{meas,k+1}\right). \tag{6.27}$$

Proof:
All nominal transitions are given in the transition function f^* according to Def. 6.3. The transition function $f^*(q_{meas,k}, e_{meas,k}^{(i)})$ is evaluated for the current measurement state $q_{meas,k}$ and the current events $e_{meas,k}^{(i)}$. If the transition function yields the following measurement state $q_{meas,k+1}$ for at least one event $e_{meas,k}^{(i)}$ the right hand side of (6.27) holds. Thus the observed transition at time k is part of the nominal transition function.
If the right hand side of (6.27) holds for the measurement sequence $k = \{1, 2, \ldots, T-1\}$, i.e. $(\forall q_{meas,k} \in \langle q_{meas,k} \rangle_{k=1}^{T-1})$, all observed state transitions are defined in the nominal transition function f^*. Thus the measurement is full transition consistent. □

Prop. 6.7 also implies that the current values of the enabler signal $w_{meas,k}$ are within the nominal limits $\boldsymbol{l}^{(i)*}$ at each switch $k = k_{\tau,i} - 1$ with $i = \{2, 3, \ldots, n_{switch}\}$.[10] Other than state consistency, full transition consistency can be achieved although a nominal transition is not triggered by the measurement data.

The notion of partial transition consistency includes specified transitions that are triggered at unexpected times.

Proposition 6.8 (Partial Transition Consistency)
The mapped state signal $\langle q_{meas,k} \rangle_{k=1}^{T-1}$, *and the set of measured events* $\mathcal{E}_{meas}$ *of the discrete event part* Z_{meas} *of a VO are called partial transition consistent with the nominal state machine* Z^*, *if*

$$\left(\exists q_{meas,k} \in \langle q_{meas,k} \rangle_{k=1}^{T-1}\right) \wedge \left(\exists \tilde{e} \in \mathcal{E}^*\right) : \left(f^*(q_{meas,k}, \tilde{e}) = q_{meas,k+1}\right) \wedge \left(\tilde{e} \neq e_{meas,k}^{(i)}\right). \tag{6.28}$$

[10] It is not necessary to check the first switch, as it represents the begin of the experiment.

Proof:
The transition function f^* is evaluated as explained in Prop. 6.7, but using the set of nominal events $\mathcal{E}^*$ instead of the set of measured events $\mathcal{E}_{meas}$. Thus $f^*(q_{meas,k}, \tilde{e})$ can yield the correct following measurement state $q_{meas,k+1}$ for any specified event $\tilde{e} \in \mathcal{E}^*$, regardless of the current measured events $e^{(i)}_{meas,k}$.
Condition (6.28) is fulfilled if there is at least one transition in the measurement sequence $k = \{1, 2, \ldots, T-1\}$ that was triggered by an unexpected event $\tilde{e} \neq e^{(i)}_{meas,k}$. □

Unspecified transitions and transitions connecting unspecified states lead to transition inconsistency according to Prop. 6.9. This represents the situation, where it is not possible to explain the observed transition by the transition function.

Proposition 6.9 (Transition Inconsistency)
The mapped state signal $\langle q_{meas,k}\rangle_{k=1}^{T-1}$, and the set of events $\mathcal{E}_{meas}$ of the discrete event part Z_{meas} of a VO are called transition inconsistent with the nominal state machine Z^*, if

$$\left(\exists q_{meas,k} \in \langle q_{meas,k}\rangle_{k=1}^{T-1}\right) \wedge \left(\forall \tilde{e} \in \mathcal{E}^*\right) \ : \ \left(f^*(q_{meas,k}, \tilde{e}) \neq q_{meas,k+1}\right). \quad (6.29)$$

Proof:
Condition (6.29) is fulfilled if there is at least one unspecified transition at a state $q_{meas,k} \in \langle q_{meas,k}\rangle_{k=1}^{T-1}$ that is not defined for any nominal event $\tilde{e} \in \mathcal{E}^*$. Therefore the measured transition is unspecified, which is inconsistent in the sense of this thesis. □

The results of state and transition consistency are combined such that the weakest result of the individual conditions determines the result of the whole discrete event system. The respective propositions are given in the following.

Proposition 6.10 (Full Discrete Consistency)
The discrete event part Z_{meas} of a VO is called full discrete consistent with the nominal state machine Z^*, if there is full state consistency according to Prop. 6.4 and full transition consistency according to Prop. 6.7.

Proof:
The main components of the discrete event part Z_{meas} according to Def. 6.4 are the set of states $\mathcal{Q}^*$, the finite set of events $\mathcal{E}^*$ and the transition function f^*. The consistency of this components is verified using the propositions about state and transition consistency. Only if all components show full consistency according to the respective propositions, the overall system also shows full consistency. Thus full discrete consistency is only given if there is full state consistency according to Prop. 6.4 and full transition consistency according to Prop. 6.7. □

Proposition 6.11 (Discrete Inconsistency)
The discrete event part Z_{meas} of a VO is called discrete inconsistent with the nominal state machine Z^*, if there is state inconsistency according to Prop. 6.6 or transition inconsistency according to Prop. 6.9.

Proof:
Discrete inconsistency is also an integral property of the main components of the discrete event part Z_{meas} as shown in Prop. 6.10. Therefore the overall discrete system is inconsistent as soon as there is at least one inconsistent component, i.e. if there is state inconsistency according to Prop. 6.6 or transition inconsistency according to Prop. 6.9. □

Proposition 6.12 (Partial Discrete Consistency)
The discrete event part Z_{meas} of a VO is called partial discrete consistent with the nominal state machine Z^*, if there is neither full discrete consistency according to Prop. 6.10 nor discrete inconsistency according to Prop. 6.11.

Proof:
If there is neither full discrete consistency according to Prop. 6.10 nor discrete inconsistency according to Prop. 6.11, there is at least one main component that shows partial consistency according to Prop. 6.5 or Prop. 6.8. Even though the other main component might show full consistency according to Prop. 6.4 or Prop. 6.7, the integral property can not be better than the properties of the included main components. □

The implementation of this propositions is straight forward, as all necessary sets and values are available in the used setting. More realistic scenarios assuming less a priori knowledge are introduced in Section 6.2 and Section 6.3.

6.1.3 Combination of the Dynamic and the Discrete Verification Results

The results of the discrete event system part can be joined with the results of the dynamic system part to achieve the overall assessment of the hybrid system. The solution of Problem 6.1 is thus given by Prop. 6.13 as follows:

Proposition 6.13 (Mapped Set of States Based Point Real Hybrid Consistency)
The mapped state signal $\mathcal{Q}_{meas}$, the interval type enclosures of the measurement data $[\boldsymbol{U}_{meas}, \boldsymbol{Y}_{meas}]$ and the set of events $\mathcal{E}_{meas}$ of the verification object H_{meas} are called consistent with a direct hybrid specification $S^*_{H,d}$, if

- *the specified state machine Z^* is full discrete consistent and*
- *the specified subsystems $s^{(i)*} \in S^*_d$ are full dynamic consistent.*

Proof:
Based on the mapped state signal according to Def. 6.10 the currently active subsystem is known at each time step. The resulting trace of the state machine is checked for consistency with the discrete event specification Z^* by regarding the states and transitions. Consistency of the states can be checked by Prop. 6.4 and consistency of the transitions by Prop. 6.7. If both parts are consistent, the measurement is consistent with the specified state machine. The dynamic of the included subsystems can be checked according to Prop. 6.2, again assuming a mapped set of measured states $\mathcal{Q}_{meas}$. If both system parts are consistent, the overall hybrid system is consistent. □

Consistency of the hybrid system results from the combination of results for the dynamic and discrete subsystems according to Prop. 6.13. Inconsistency of the hybrid system is given as soon as one subsystem is inconsistent. Else the hybrid system is partial consistent, meaning that there is no inconsistent subsystem but at least on subsystem is partial consistent.

An overview of the different propositions and possible results is given in Tab. 6.1.

Table 6.1: Consistency criteria

Part of the System	Property: full consistency	Property: partial consistency	Property: inconsistency
Dynamic Subsystems	Prop. 6.2	not applicable	Prop. 6.3
States	Prop. 6.4	Prop. 6.5	Prop. 6.6
Transitions	Prop. 6.7	Prop. 6.8	Prop. 6.9
State Machine	Prop. 6.10	Prop. 6.12	Prop. 6.11
Hybrid System	Prop. 6.13	else	Prop. 6.3 or Prop. 6.11

By applying Prop. 6.2, consistency of the dynamic part of the hybrid system can be shown in a straight forward way. The results are calculated based on the propositions of the previous chapters and thus show the same guaranteed properties as defined there. A direct specification $\mathcal{S}_d^*$ was used throughout the chapter for notational simplicity. It is straight forward to extend all propositions to hold also for an interval type specification $\mathcal{S}_i^*$. This is due to the fact that the dynamic verification is based on the united solution set given by the measurement data. Therefore all properties stay the same except that the verification of dynamic consistency is done based on Prop. 5.1 instead of Prop. 4.1.

Subsystems that are verified using the Kaucher based approach are guaranteed to be correct and it is not possible that there are any hidden faults present in the system.

The state machine has to use point real numbers in both cases. Therefore the definitions and propositions for the discrete event part are unchanged.

An example of the hybrid verification procedure for a mapped set of measured states is given in Example 6.1.

Example 6.1:

Assume a direct hybrid specification $S^*_{H,d} = [Z^*, S^*_d]$. *The nominal state machine* Z^* *is depicted in Fig. 6.3.*

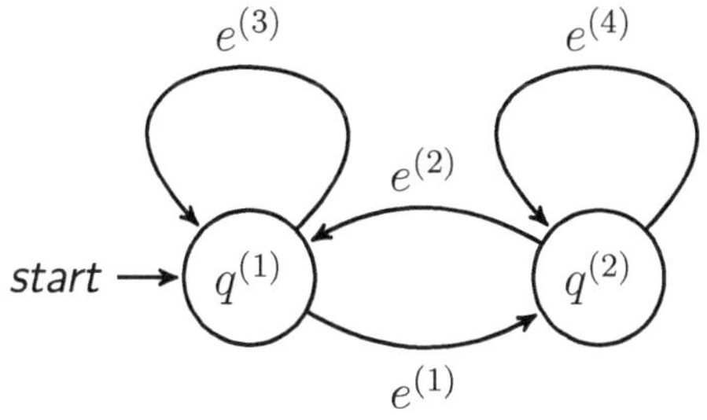

Figure 6.3: Specified state machine Z^*

The state machine consists of two states $\mathcal{Q}^* = \{q^{(1)*}, q^{(2)*}\} = \{1, 2\}$, *the events are based on the enabler signal that is defined to be the output signal* $w = y$:

$$e^{(1)*} : w_k \in \boldsymbol{l}^{(1)} = [1,\ 2] \tag{6.30}$$

$$e^{(2)*} : w_k \in \boldsymbol{l}^{(2)} = [31,\ 33] \tag{6.31}$$

$$e^{(3)*} : w_k \in \boldsymbol{l}^{(3)} = [-\infty,\ \infty] \tag{6.32}$$

$$e^{(4)*} : w_k \in \boldsymbol{l}^{(4)} = [-\infty,\ \infty]\,. \tag{6.33}$$

The events $e^{(3)*}$ *and* $e^{(4)*}$ *are enabled for all values of the enabler signal, leading to the permanent possibility to stay in the current state, based on the transition function:*

$$f^*\left(q^{(1)*}, e^{(1)*}\right) = q^{(2)*} \tag{6.34}$$

$$f^*\left(q^{(2)*}, e^{(2)*}\right) = q^{(1)*} \tag{6.35}$$

$$f^*\left(q^{(1)*}, e^{(3)*}\right) = q^{(1)*} \tag{6.36}$$

$$f^*\left(q^{(2)*}, e^{(4)*}\right) = q^{(2)*}. \tag{6.37}$$

The set of dynamic systems S^*_d *is given by first order systems, i.e.* $n_a(i) = n_c(i) = 1$, $\forall\, i \in \{1, 2\}$ *leading to*

$$s^{(i)*} : y_k = a^*_1(i) y_{k-1} + c^*_1(i) u_{k-1}. \tag{6.38}$$

The nominal parameters of the dynamic subsystems are given in Tab. 6.2.

Table 6.2: Nominal parameters of the subsystems $s^{(1)*}$ and $s^{(2)*}$

Subsystem	a^*_1	c^*_1
$q^{(1)*}$	0.1	1
$q^{(2)*}$	2.0	1

There are no optional initial input and output values given. Thus the input and output values are kept across the switches.
The implementation is assumed to be done by one or more human developers. Therefore there might be inconsistencies in the resulting VO. Note that the implemented system is assumed to consist of real hard- and software and to include a given plant that cannot be changed. Therefore the corresponding state machine Z_{meas} *and its dynamical subsystems* $\mathcal{S}_{meas}$ *are not directly known. Nevertheless it is possible to excite the system and measure its output and state signals. The random excitation signal used in this example is given by*

$$u_k = 1 + 0.2\eta_k, \tag{6.39}$$

where η_k *is drawn from a standard normal distribution. The resulting measurement data is given in Fig. 6.4. Thereby the output was enclosed by intervals using an additive fault of* $\delta_y^a = 0.5$. *The switching times are based on the information given in the mapped state signal*

$$Q_{meas} = [1, 1, 1, 1, 1, 2, 2, 2, 2, 1, 1, 1, 1, 1, 1] \tag{6.40}$$

leading to the switches $k_\tau = [1, 6, 10]$. *The relevant values of the enabler signal are given at the time steps right before a switch* $\boldsymbol{w}_{k_{\tau,i}-1}$ *with* $i \in \{2, 3\}$ *i.e.* $\boldsymbol{w}_5 = [0.6,\ 1.1]$ *and* $\boldsymbol{w}_9 = [31.9,\ 32.9]$.

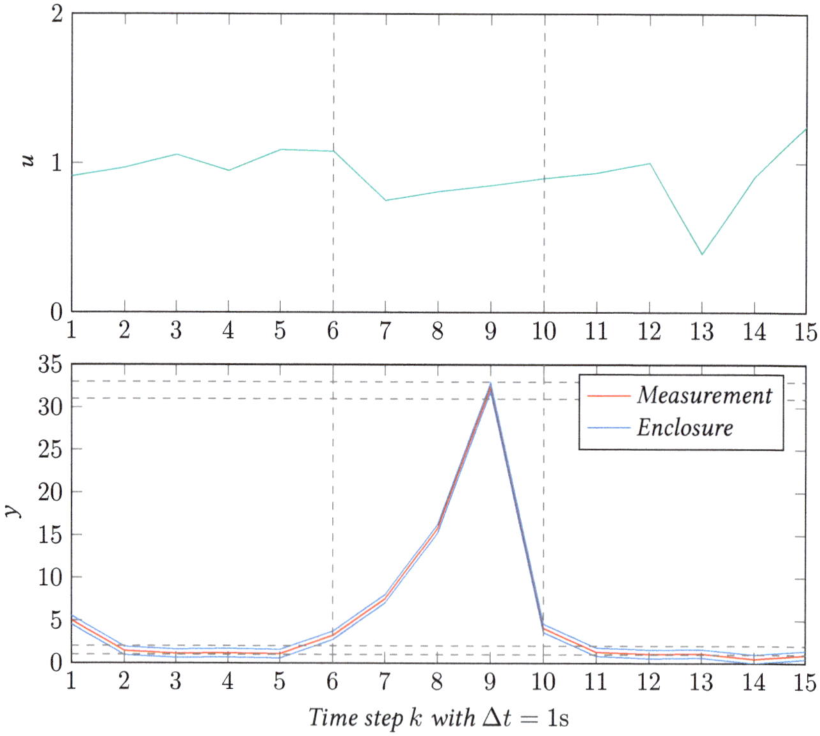

Figure 6.4: Measured trajectory and subsystem switches

Dynamic Consistency

The subsystems are verified using the introduced Kaucher based method. The resulting feasibility signals are depicted in Fig. 6.5. The first verification result can be calculated for $k_{min} = \max(n_a, n_c) + 1$*. Due to the autoregressive system of order* $n_a = n_c = 1$ *in this example it is thus not possible to calculate a verification result for the very first element of each segment. It can be seen, that all three segments* $j = \{1, 2, 3\}$ *can be explained by the respective mapped nominal states* $i = \{1, 2\}$ *using the Kaucher based method according to Prop. 6.2. Therefore the dynamic subsystems* $\mathcal{S}_{meas}$ *are full consistent with the specification* $\mathcal{S}_d^*$*.*

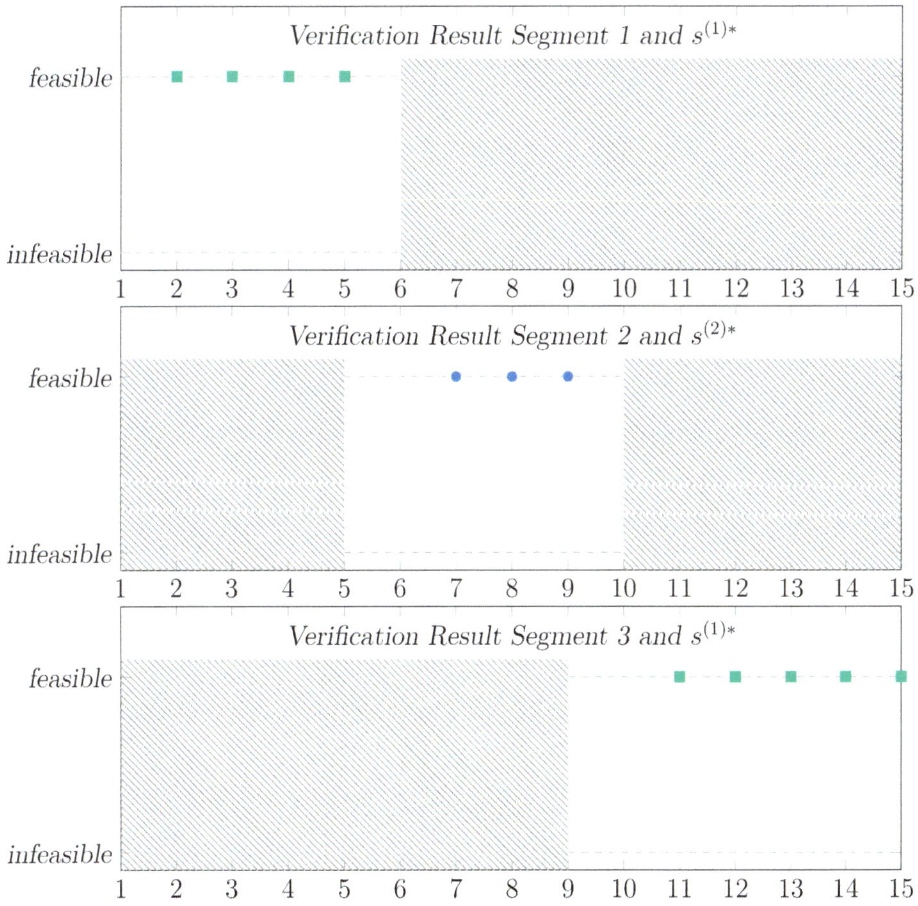

Figure 6.5: Verification result for each segment (using a mapped state signal)

Discrete Consistency

There is a mapped set of states given in this example. Therefore the measured set of states is given by the unique values in the mapped state signal Q_{meas}

$$\mathcal{Q}_{meas} = \{1, 2\} = \left\{q^{(1)*}, q^{(2)*}\right\} = \mathcal{Q}^* \tag{6.41}$$

which leads to full state consistency according to Prop. 6.4.
The set of measured events can be extracted from the measurement data. First, the events leading to the switches in $k \in \{6, 10\}$ are determined:

$$e^{(1)}_{meas,5} : \boldsymbol{w}_5 = [0.6,\ 1.6] \cap [1,\ 2] \neq \emptyset : e^{(1)*} \tag{6.42}$$
$$e^{(2)}_{meas,5} : \boldsymbol{w}_5 = [0.6,\ 1.6] \cap [-\infty,\ \infty] \neq \emptyset : e^{(3)*} \tag{6.43}$$
$$e^{(3)}_{meas,5} : \boldsymbol{w}_5 = [0.6,\ 1.6] \cap [-\infty,\ \infty] \neq \emptyset : e^{(4)*} \tag{6.44}$$
$$e^{(1)}_{meas,9} : \boldsymbol{w}_9 = [31.9,\ 32.9] \cap [31,\ 33] \neq \emptyset : e^{(2)*} \tag{6.45}$$
$$e^{(2)}_{meas,9} : \boldsymbol{w}_9 = [31.9,\ 32.9] \cap [-\infty,\ \infty] \neq \emptyset : e^{(3)*} \tag{6.46}$$
$$e^{(3)}_{meas,9} : \boldsymbol{w}_9 = [31.9,\ 32.9] \cap [-\infty,\ \infty] \neq \emptyset : e^{(4)*}\ . \tag{6.47}$$

This means that $\left\{e^{(1)*}, e^{(3)*}, e^{(4)*}\right\}$ are activated at $k = 5$ and $\left\{e^{(2)*}, e^{(3)*}, e^{(4)*}\right\}$ are activated at $k = 9$. The events can now be applied to the nominal transition function:

$$f^*\left(q_{meas,5}, e^{(1)}_{meas,5}\right) = f^*\left(q^{(1)*}, e^{(1)*}\right) \stackrel{!}{=} q^{(2)*} = q_{meas,6} \tag{6.48}$$
$$f^*\left(q_{meas,5}, e^{(2)}_{meas,5}\right) = f^*\left(q^{(1)*}, e^{(3)*}\right) \stackrel{!}{=} q^{(1)*} \neq q_{meas,6} \tag{6.49}$$
$$f^*\left(q_{meas,5}, e^{(3)}_{meas,5}\right) = f^*\left(q^{(1)*}, e^{(4)*}\right) = \emptyset \tag{6.50}$$
$$f^*\left(q_{meas,9}, e^{(1)}_{meas,9}\right) = f^*\left(q^{(2)*}, e^{(2)*}\right) \stackrel{!}{=} q^{(1)*} = q_{meas,10} \tag{6.51}$$
$$f^*\left(q_{meas,9}, e^{(2)}_{meas,9}\right) = f^*\left(q^{(2)*}, e^{(3)*}\right) = \emptyset \tag{6.52}$$
$$f^*\left(q_{meas,9}, e^{(3)}_{meas,9}\right) = f^*\left(q^{(2)*}, e^{(4)*}\right) \stackrel{!}{=} q^{(2)*} \neq q_{meas,10}. \tag{6.53}$$

It can be seen that (6.48) and (6.51) hold. This leads to the verification of the transition function in $k \in \{5, 9\}$, that generates the switches $k_\tau \in \{6, 10\}$. The events $e^{(3)*}$ and $e^{(4)*}$ are enabled for all $k \in \{1, 2, \ldots, 15\}$ and provide the possibility to stay in the same state for several time steps. To improve readability, only the verification of one exemplary step is shown. The activated events are

$$e^{(1)}_{meas,8} : \boldsymbol{w}_8 = [15.3,\ 16.3] \cap [-\infty,\ \infty] \neq \emptyset : e^{(3)*} \tag{6.54}$$
$$e^{(2)}_{meas,8} : \boldsymbol{w}_8 = [15.3,\ 16.3] \cap [-\infty,\ \infty] \neq \emptyset : e^{(4)*}. \tag{6.55}$$

These are applied to the nominal transition function

$$f^*\left(q_{meas,8}, e^{(1)}_{meas,8}\right) = f^*\left(q^{(2)*}, e^{(3)*}\right) = \emptyset \tag{6.56}$$
$$f^*\left(q_{meas,8}, e^{(2)}_{meas,8}\right) = f^*\left(q^{(2)*}, e^{(4)*}\right) \stackrel{!}{=} q^{(2)*} = q_{meas,9}. \tag{6.57}$$

Thus the behavior is valid at time $k = 8$. Similar results are obtained for all other time steps. The nominal transition function holds $\forall q_{meas,k} \in Q_{meas}$, which leads to full transition consistency according to Prop. 6.7. Hence full state consistency (Prop. 6.4) and full transition consistency (Prop. 6.7) hold in the given example. This leads to full discrete consistency between Z_{meas} and Z^* according to Prop. 6.10.

Hybrid Consistency

The previous partial results are now combined with respect to hybrid consistency as given in Prop. 6.13. It was shown that Z_{meas} is full consistent with Z^* and that $\mathcal{S}_{meas}$ is full consistent with $\mathcal{S}_d^*$. Therefore the verification object H_{meas} and the specification $S_{H,d}^*$ are consistent. This means that the superimposed state machine as well as the linear dynamic subsystems of the VO that produced the measurement in Fig. 6.4 are full consistent with the specification of $S_{H,d}^*$ given in this example. Therefore the VO is verified with respect to the nominal system.

6.2 Verification of Hybrid Systems With Given Switching Times

In general it is not possible to measure the internal signals and states of a VO. Therefore the setting is changed and the assumption of an available mapped state signal Q_{meas} is dropped. Nevertheless, it is still assumed that the correct times of the switches k_τ are available, even though the respective active states are unknown. The resulting consistency problem is given in the following:

Problem 6.2 (Point Real Hybrid Consistency with Given Switches)
Is the nominal hybrid system, specified by a direct hybrid specification

$$S^*_{H,d} = \{Z^*, S^*_d\} \tag{6.58}$$

consistent with the input-output behavior given by the interval type enclosures of T measurement values

$$[\boldsymbol{U}_{meas}, \mathbf{Y}_{meas}] = \left[\langle \boldsymbol{u}_{meas,k}\rangle_{k=1}^{T}, \langle \boldsymbol{y}_{meas,k}\rangle_{k=1}^{T}\right] \tag{6.59}$$

and the set of switches

$$\{k_{\tau,j}\}_{j=1}^{n_{switch}}, \tag{6.60}$$

i.e. can the measurement data be explained by the nominal system?

To solve the problem, it is necessary to determine a mapped state signal Q. The measurement data can be segmented based on the given switches. The result can be interpreted as an unmapped state signal as it is not known which state is active after each switch. Therefore is is necessary to determine the correct nominal subsystem for each segment. There are two important assumptions that need to hold to ensure an unambiguous mapping from nominal subsystems to measurement segments.

Assumption 6.2 (Prager-Oettli-Distinguishability of the Set of Subsystems)
Each two nominal subsystems $s^{(i)*}, s^{(j)*} \in S^*_d$ with $i \neq j$ are called Prager-Oettli-Distinguishable (PO-Distinguishable) with respect to specific segmented measurement data $\left\langle \left[\mathbf{U}^{(j)}_{meas}, \mathbf{Y}^{(j)}_{meas}\right]\right\rangle_{j=1}^{n_{switch}}$, if there is no segment $j \in \{1, 2, \ldots, n_{switch}\}$ that fulfills Prop. 4.1 for both subsystems $s^{(i)*}$ and $s^{(j)*}$.

This means that two distinct nominal subsystems given in the specification are sufficiently different with respect to the measurement data, noise assumptions and the interval enclosure. The second assumption transfers this property to the segments of the measurement data.

Assumption 6.3 (Mappability of the Measurement)
The segmented measurement data $\left\langle \left[\mathbf{U}^{(j)}_{meas}, \mathbf{Y}^{(j)}_{meas}\right] \right\rangle_{j=1}^{n_{switch}}$ *is called mappable to the set of dynamic subsystems* $\mathcal{S}^*_d$*, if there is no specified subsystem* $s^{(i)*} \in \mathcal{S}^*_d$ *that fulfills Prop. 4.1 for any segment* $\left[\mathbf{U}^{(j)}_{meas}, \mathbf{Y}^{(j)}_{meas}\right]$ *that was generated by another dynamic system* $s^{(j)}$ *with* $s^{(j)} \neq s^{(i)*}$.

If Assumption 6.2 and 6.3 hold, it is possible to determine a mapped state signal. Therefore Problem 6.2 can be solved by the following proposition:

Proposition 6.14 (Point Real Hybrid Consistency with Given Switches)
The direct hybrid specification $S^*_{H,d}$*, the interval type enclosures of the measurement data* $[\boldsymbol{U}_{meas}, \boldsymbol{Y}_{meas}]$ *and the set of switches* $\{k_{\mathcal{T},j}\}_{j=1}^{n_{switch}}$ *of a verification object* H_{meas} *can be used to set up a mapped state signal* Q *if the specified set of subsystems* S^*_d *is PO-distinguishable with respect to the measurement data according to Assumption 6.2 and the measurement data is mappable to* S^*_d *according to Assumption 6.3.*
The availability of a mapped state signal Q *transforms the problem to a mapped set of states based point real hybrid consistency problem. This problem can be solved using Prop. 6.13.*

Proof:
An unmapped state signal can directly be constructed from the given correct switches $\{k_{\mathcal{T},j}\}_{j=1}^{n_{switch}}$ according to (6.19). As Assumption 6.2 and 6.3 hold, individual nominal subsystems within the specification $\mathcal{S}^*_d$ can be distinguished from each other.
Also, the united solution set $\sum^{(j)}_{\exists\exists}$ defined by the measurement data of each segment $\left[\mathbf{U}^{(j)}_{meas}, \mathbf{Y}^{(j)}_{meas}\right]$, generated by a dynamic system $s^{(j)}$, cannot be explained by any other nominal system $s^{(i)*} \in \mathcal{S}^*_d$, with $s^{(j)} \neq s^{(i)*}$. The generating subsystem can thus be determined unambiguously for each segment. The mapped nominal subsystems can be used to determine the active states q_k with $k \in \{k_{\mathcal{T},j}, k_{\mathcal{T},j}+1, \ldots, k_{\mathcal{T}',j}\}$ for all segments $j \in \{1, 2, \ldots, n_{switch}\}$. The resulting signal Q is a mapped state signal according to Def. 6.10. The given measurement data $[\boldsymbol{U}_{meas}, \boldsymbol{Y}_{meas}]$, together with the given set of nominal subsystems $\mathcal{S}^*_d$ and the extracted mapped state signal Q represents the setting of Problem 6.1 that can be solved by Prop. 6.13. □

Segments generated by an unspecified subsystem $\tilde{s} \notin \mathcal{S}_d^*$ cannot be mapped to a nominal subsystem $s^{(i)*}$ if $\tilde{s}$ is PO-Distinguishable from all nominal subsystems $s^{(i)*} \in \mathcal{S}_d^*$ according to Assumption 6.2. If this is not the case, the measurement data generated by $\tilde{s}$ can be explained by at least one nominal subsystem $s^{(i)*} \in \mathcal{S}_d^*$ and thus it is impossible to recognize the subsystem $\tilde{s}$. Note that the definition is based on all measurement data of a segment, i.e. it is possible that partial measurement data of a segment can be included in more then one nominal subsystem.

The method leads to a mapping algorithm that compares the dynamic of each nominal subsystem, given by the n_q states, with every segment $j \in \{1, 2, \ldots, n_{switch}\}$ of the measurement data. This comparison is done based on the united solution set, as given in Prop. 4.1. The flowchart of the algorithm is given in Fig. 6.6.
An exemplary application of point real hybrid consistency with known switches is given in Example 6.2.

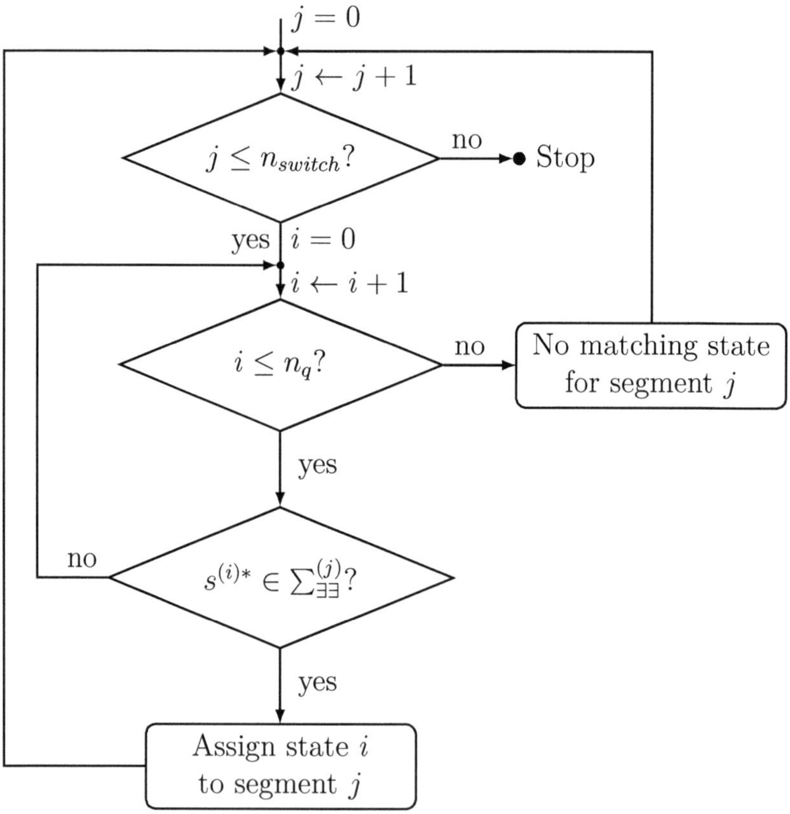

Figure 6.6: Flowchart of the mapping algorithm

Example 6.2:

Assume the same setting as introduced in Example 6.1, without the assumption of a given measured mapped state signal. Instead, a set of switches is given by

$$k_{\mathcal{T}} \in \{1, 6, 10\} . \tag{6.61}$$

Based on the switches and the length $T = 15$ of the measurement data, the endpoints of the segments can be calculated:

$$k_{\mathcal{T}'} \in \{5, 9, 15\} . \tag{6.62}$$

The nominal parameters given in Tab. 6.2 and the algorithm of Fig. 6.6 are used to determine the mapping between segments and states.
To show the correct classification of all subsystems, the algorithm is altered such that it compares all possible nominal subsystems with each segment instead of proceeding to the next segment as soon as a consistent subsystem is found.
The results are depicted in Fig. 6.7. Each subfigure shows the evaluation of both nominal subsystems $s^{(1)*}$ and $s^{(2)*}$ for one of the segments $j \in \{1, 2, 3\}$. The shaded areas mark measurement data that does not belong to the segments and thus is not taken into account for the respective verification. It can be seen, that only one generating nominal subsystem can be verified for each of the three segments. This leads to an unambiguous mapping

$$Q = [1, 1, 1, 1, 1, 2, 2, 2, 2, 1, 1, 1, 1, 1, 1] \tag{6.63}$$

which is the same as given in (6.40).
The matching algorithm is based on Prop. 6.1. Therefore dynamic consistency is given, if it is possible to generate a mapped state signal Q. The discrete verification can be done as described in the previous section. The combination of both yields the hybrid verification result. Both results are equivalent with the calculations and results demonstrated in Example 6.1 and are thus not repeated here.

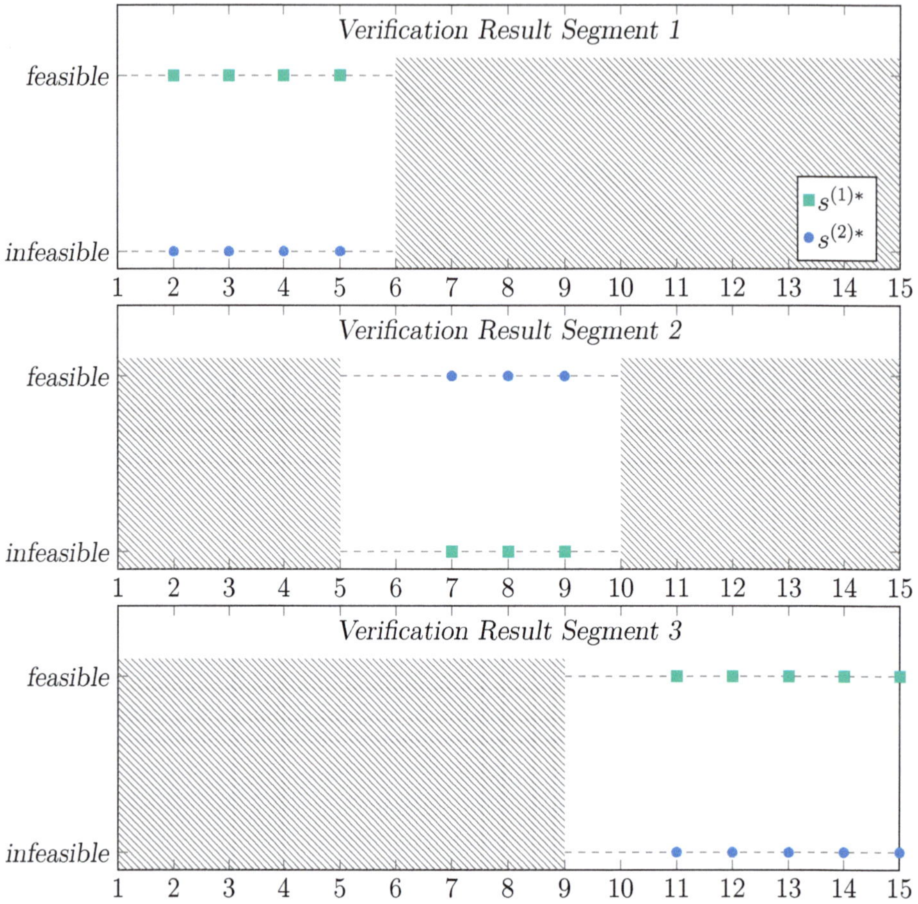

Figure 6.7: Verification Result for each segment (without mapped set of states)

6.3 Verification of Hybrid Systems With Unknown Switching Times

In the last step even the knowledge about the switching times is dropped. Only the specification and the input and output measurement data are available. Thus the switching times as well as the respective modes need to be reconstructed from the measurement data only. Therefore an additional segmentation step is necessary in the procedure. This step aims at finding the unknown switches k_τ, at which the active generating subsystem changes. The corresponding verification problem is given in Problem 6.3.

Problem 6.3 (Point Real Hybrid Consistency)
Is the nominal hybrid system, specified by a direct hybrid specification

$$S^*_{H,d} = \{Z^*, \mathcal{S}^*_d\} \tag{6.64}$$

consistent with the input-output behavior given by the interval type enclosures of T measurement values

$$[\boldsymbol{U}_{meas}, \boldsymbol{Y}_{meas}] = \left[\langle \boldsymbol{u}_{meas,k}\rangle_{k=1}^{T}, \langle \boldsymbol{y}_{meas,k}\rangle_{k=1}^{T}\right] \tag{6.65}$$

without any knowledge about the state or the switching times?

The problem can be solved, if it is possible to determine a mapped state signal from the input-output data only. If this signal is constructed, Problem 6.3 can again be reduced to Problem 6.1 which is solved by Prop. 6.13. However, in this setting there is no information about the active states or the switches. Therefore an additional identification and segmentation procedure is necessary. The procedure introduced in this thesis is based on previous work of the author, published in [Die13a], [Die13b] and used in [Die17]. However, while the original work aims on identifying an a priori unknown library of subsystems from the measurement data, the concept is adapted in this thesis to consider a given set of nominal subsystems. Additionally, it is extended to the interval arithmetic context of guaranteed verification.
First, switches that are detected by the segmentation and identification method are defined.

Definition 6.11 (Detected Switch)
The time instant k that does not show dynamic consistency according to Prop. 6.1 of any nominal subsystems $s^{(i)*} \in \mathcal{S}^*_d$ with the current regressor matrix $\boldsymbol{A}_{meas,k}$ and the current measurement vector $\boldsymbol{B}_{meas,k}$ is called detected switch $\hat{k}_{\tau,j+1} = k$.

To ensure correct segmentation, the available measurement data needs to be Prager-Oettli-Segmentable:

Assumption 6.4 (Prager-Oettli-Segmentability)
The measurement data $\left[\mathbf{U}^{(j)}_{meas}, \mathbf{Y}^{(j)}_{meas}\right]$ and $\left[\mathbf{U}^{(j+1)}_{meas}, \mathbf{Y}^{(j+1)}_{meas}\right]$ of two consecutive segments j and $j+1$ that are generated by distinct subsystems $s^{(j)*} \neq s^{(j+1)*}$ are called Prager-Oettli-Segmentable with respect to a given set of subsystems $\mathcal{S}^*_d$, if there is no subsystem $s^{(i)*} \in \mathcal{S}^*_d$ that fulfills Prop. 6.1 for all measurement data points in the combined segment $[k_{\tau,j}, k_{\tau',j+1}]$.

Assumption 6.4 implies Prager-Oettli-Distinguishability of the set of subsystems $\mathcal{S}^*_d$ according to Assumption 6.2. Based on the segmentability assumption it is possible to extract a mapped state signal to solve Problem 6.3.

Proposition 6.15 (Point Real Hybrid Consistency)
The direct hybrid specification $S^*_{H,d}$ and the interval type enclosures of the measurement data $[\boldsymbol{U}_{meas}, \boldsymbol{Y}_{meas}]$ of a verification object H_{meas} can be used to set up a mapped state signal Q, if the measurement data is PO-segmentable with respect to $S^*_{H,d}$.
The availability of a mapped state signal Q transforms the problem to a mapped set of states based point real hybrid consistency problem. This problem can be solved using Prop. 6.13.

Proof:
If Prager-Oettli-Distinguishability (Assumption 6.2) and mappability of the measurement (Assumption 6.3) are fulfilled, there is only one possible consistent subsystem $s^{(i)*} \in \mathcal{S}^*_d$ at the end of a segment $\hat{k}_{\mathcal{T}',j} = \hat{k}_{\mathcal{T},j+1} - 1$. This subsystem needs to represent the active state of the whole segment $\left[\hat{k}_{\mathcal{T},j}, \hat{k}_{\mathcal{T}',j}\right]$.
On the other hand, it is impossible that this subsystem is consistent with the measurement data of the entire next segment generated by a different subsystem $s^{(j+1)}$. A nominal subsystem $s^{(i)*} \in \mathcal{S}^*_d$ that fulfills Prop. 6.1 for all measurement points in the entire combined segment $[k_{\mathcal{T},j}, k_{\mathcal{T}',j+1}]$ will also fulfill Prop. 6.1 for any part of the combined segment. This holds especially for the parts $[k_{\mathcal{T},j}, k_{\mathcal{T}',j}]$ and $[k_{\mathcal{T},j+1}, k_{\mathcal{T}',j+1}]$, i.e. the same nominal subsystem is consistent with two distinct segments $\left[\mathbf{U}^{(j)}_{meas}, \mathbf{Y}^{(j)}_{meas}\right]$ and $\left[\mathbf{U}^{(j+1)}_{meas}, \mathbf{Y}^{(j+1)}_{meas}\right]$. This is only possible if $s^{(i)*} = s^{(j)} = s^{(j+1)}$ which is not allowed in the case of mappability of the measurement according to Assumption 6.3.
Thus it is possible to determine a switch $\hat{k}_{\mathcal{T},j} \geq k_{\mathcal{T},j}$ for each $j \in \{2, 3, \ldots, n_{switch}\}$, i.e. the true number of segments is determined even if the detected switches do not exactly match the real ones. Based in this result it is possible to determine the respective active subsystem for all segments, whereas each estimated segment $\left[\hat{k}_{\mathcal{T},j},\ \hat{k}_{\mathcal{T}',j}\right]$ at least partly overlaps with the true segment $[k_{\mathcal{T},j},\ k_{\mathcal{T}',j}]$. This leads to a mapped state signal with correct mapping for all segments, even if the detected segment boundaries might slightly differ from the true ones.[11]
Thus Problem 6.3 is transformed to Problem 6.1 which completes the proof. □

If Prager-Oettli-Segmentability according to Assumption 6.4 does not hold, it is not possible to determine the switch. The flowchart of the algorithm implementing the introduced identification and segmentation method is given in Fig. 6.8.

[11] A detailed proof of this property is given in Section 6.3.1.

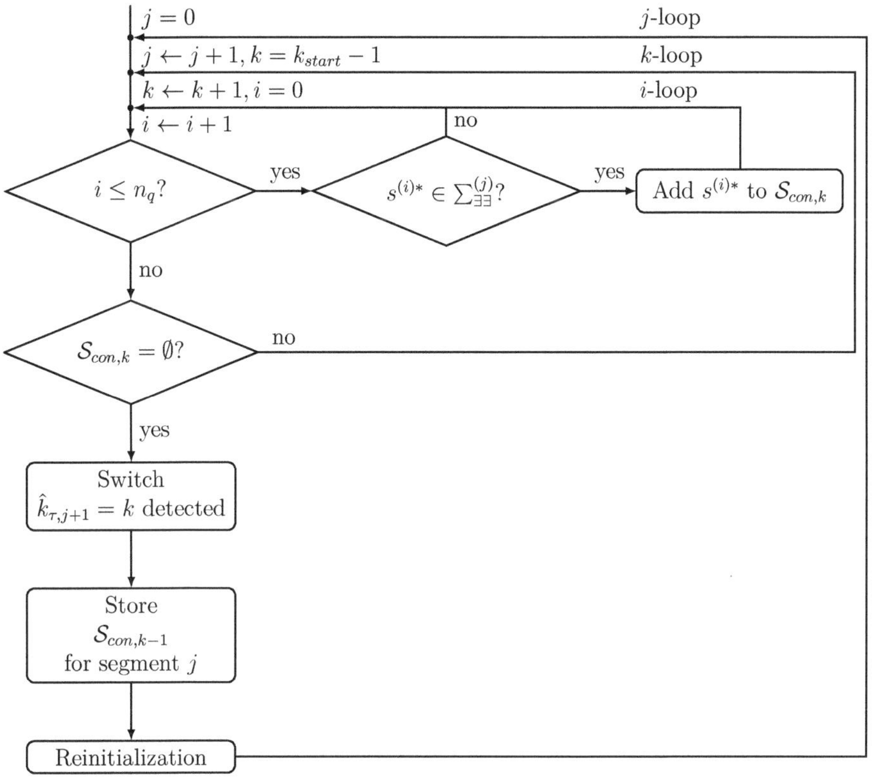

Figure 6.8: Flowchart of the identification and segmentation algorithm

The inner loop ('i-loop') depicted in the flow chart compares all $n_q = |\mathcal{S}_d^*|$ nominal subsystems $s^{(i)*} \in \mathcal{S}_d^*$, with the current segment. If there is a consistent nominal subsystem, it is added to the set of consistent subsystems $\mathcal{S}_{con}$. The next loop ('k-loop') is running as long as the set of consistent subsystems $\mathcal{S}_{con,k}$ is non-empty. If this is the case, the current segment can be explained by at least one nominal subsystem. Therefore the segment is extended by one time step.

This new segment is again verified by the i-loop. If the set of consistent subsystems is empty, i.e. $\mathcal{S}_{con} = \emptyset$, none of the nominal subsystems is able to explain the current segment. However, the measurement was verified in the previous step. Therefore a detected switch[12] is recognized at $\hat{k}_{\tau,j+1} = k$ and the active subsystems for the segment j are given by $\mathcal{S}_{con,k-1}$. Note that due to Assumption 6.3 only one subsystem is allowed to explain an entire segment of the measurement data. This leads to $\left|\mathcal{S}_{con,\hat{k}_{\tau',j}}\right| \overset{!}{=} 1$.

However, multiple consistent subsystems $|\mathcal{S}_{con,k}| \geq 1$ are possible for partial segments $k \in \{k_{\tau,j}+1, k_{\tau,j}+2, \ldots, k_{\tau',j}-1\}$.

Detecting a switch and determining the state of the finished segment ends the k-loop of the algorithm in Fig. 6.8. The measurement values belonging to the just finished segment are removed from the considered measurement data.

12 The first switch of a system is defined to be $\hat{k}_{\tau,1} = 1$, according to Def. 6.7ff.

All counters and intermediate values are reinitialized and the next iteration of the outer loop ('j-loop') starts for the following segment.

In order to achieve the hybrid verification result, the discrete and dynamic results are combined according to Section 6.1.3. An application of the procedure is given in Example 6.3.

Example 6.3:

Assume the same setting as introduced in Example 6.1 except that there are neither a mapped state signal nor any information about the switches.
The first iteration of the j-loop of the identification and segmentation algorithm (Fig. 6.8) is depicted in the first subplot of Fig. 6.9.

Both subsystems are considered for verification in each step of the k-loop. It is possible to verify subsystem $s^{(1)*}$ for $k \in \{2, 3, 4, 5\}$. The first result can again be calculated for $k = 2$ and the first switch is recognized at $k_{\tau,2} = 6$. This is the first time both dynamic subsystems show inconsistency, i.e. $\mathcal{S}_{con,6} = \emptyset$.
Note that the algorithm will break the k-loop at $k = 6$. However this was not done in Fig. 6.9 to show that the verification results stay infeasible for all regarded segments in the remaining measurement time.
In the reinitialization step, the detected first segment is deleted from the measurement data. This is depicted as shaded area in subfigure 2 and 3 of Fig. 6.9. The second iteration of the j-loop verifies subsystem $s^{(2)*}$ for $k \in \{7, 8, 9\}$ and detects the next switch at $k_{\tau,3} = 10$. Subsystem $s^{(1)*}$ is evaluated to be infeasible for the whole available measurement. Again the infeasible result for both subsystems are given until the end of the measurement in contrary to the genuine break of the k-loop.
The third segment shows consistency with subsystem $s^{(1)*}$ for $k \in \{11, 12, 13, 14, 15\}$ which leads to $k_{\tau',3} = T$. This is feasible with respect to Def. 6.7ff. The subsystem $s^{(2)*}$ is evaluated to be infeasible for segment 3.
The results can be used to set up the matched state signal

$$Q = [1, 1, 1, 1, 1, 2, 2, 2, 2, 1, 1, 1, 1, 1, 1] \tag{6.66}$$

which corresponds again with the ground truth given in (6.40). The successful mapping for all time steps implies continuous consistency as there is an unambiguous nominal subsystem mapped to each time step.

The nominal state machine Z^* of Example 6.1 is used here as well. Due to $Q = Q_{meas}$, all values needed for the verification of the discrete part are the same as in Example 6.1. This leads to full discrete consistency between the unsegmented measurement data and the specification.
Finally, full hybrid consistency can also be concluded for this setting. This shows a successful verification of a hybrid system only based on interval type measurements of the input-output data and the nominal system $S^*_{H,d}$.

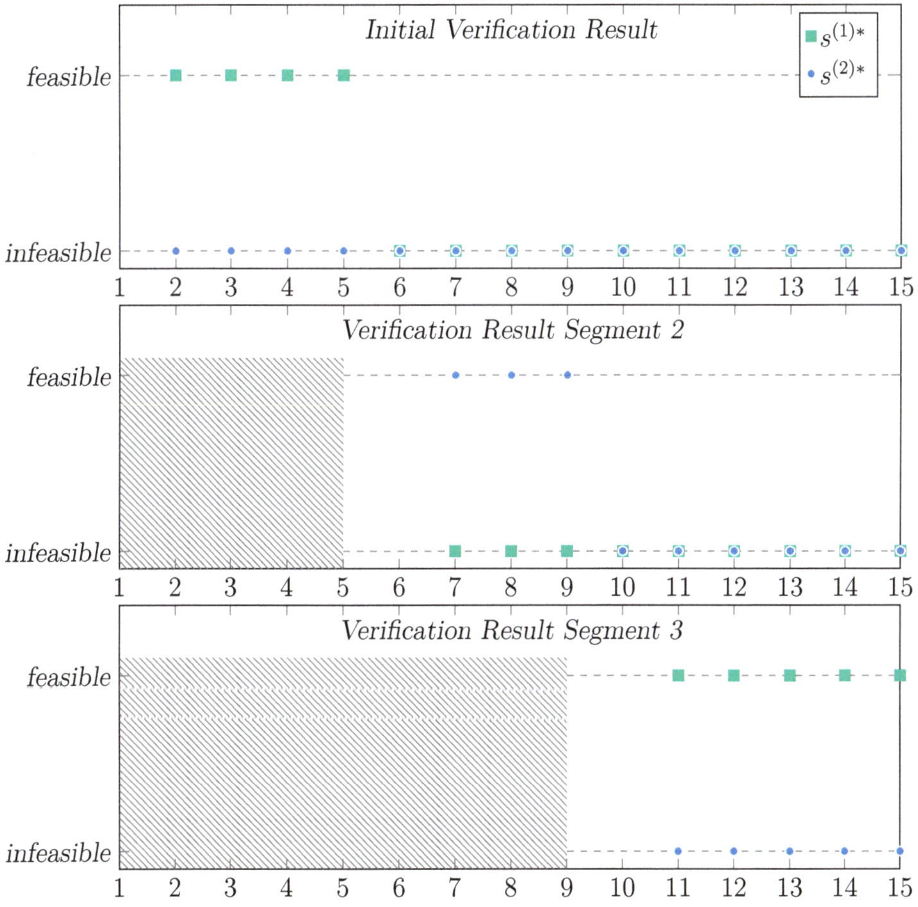

Figure 6.9: Verification result for each segment (without segmentation)

6.3.1 Convergence of the Identification and Segmentation Algorithm

To show convergence of the identification and segmentation algorithm it is possible to extract more information about the switches from the measurement data by adding a backward iteration. As a basic property of the identification and segmentation algorithm the true switches are always overestimated and never underestimated. This can be shown by considering extreme values of the additive noise ϵ as follows:

Noise tending to zero ($\epsilon \to 0$) When the noise tends to zero, the interval enclosure necessary to bound the noise tends to zero, too. Prop. 4.1 has thus to be fulfilled for each line of the regressor matrix individually without any deviations. If there is at least one measurement point added to the regressor matrix that is generated by a different and Prager-Oettli-Distinguishable subsystem, the data is not consistent anymore. The switch can thus be detected at the first time step of the new interval, i.e. at the true switch with $\hat{k}_{\mathcal{T},j} = k_{\mathcal{T},j}$.

Noise tending to infinity ($\epsilon \to \infty$) When the noise tends to infinity, the enclosing interval width also tends to infinity to contain the noisy data. Therefore the regressor matrix provides as well infinite possible entries to fulfill Prop. 4.1 which leads to consistency for any measurement data. Hence it is ensured that for the detected switch holds $\hat{k}_{\mathcal{T},j} \geq k_{\mathcal{T},j}$.

Note that $\epsilon \to \infty$ will also lead to the violation of Prager-Oettli-Segmentability given in Assumption 6.4, as well as the violation of the full rank Assumption 3.1. In practice it is thus important to ensure a suitable ϵ such that all assumptions of distinguishability, segmentability and the rank are met.
It is not possible to determine the precise amount of time $k_{over} = \hat{k}_{\mathcal{T},j} - k_{\mathcal{T},j}$ the switch will be overestimated. Nevertheless there will be a gradient from instantaneous detection $k_{over} = 0$ in the case of $\epsilon \to 0$ and the maximum overestimation in the case of $\epsilon \to \infty$.

To determine the switch more precisely, the detection algorithm can be extended with an antichronological iteration. The end time of the last segment is thereby given as the time of the last measurement $k_{\mathcal{T}',n_{switch}} = T$ by definition. The regressor matrix is set up such that it includes the values from a variable time k_{start} up to the final time T. If the dynamic consistency of Prop. 4.1 is fulfilled for all measurement values $\left[\langle \mathbf{u}_{meas,k}\rangle_{k=k_{start}}^{T} \; \langle \mathbf{y}_{meas,k}\rangle_{k=k_{start}}^{T}\right]$, the segment is extended backwards in time by reducing $k_{start} \leftarrow k_{start} - 1$. If the segment extends over a switch, i.e. $k_{start} < k_{\mathcal{T},j}$, data from two distinct subsystems is included in the regressor matrix. The detection of this switch is dependent on the same assumptions as in forward direction, namely Prager-Oettli-Distinguishability and Prager-Oettli-Segmentability. The switches detected backwards in time show the same behavior of overestimation as the switches detected in the forward iteration.

The initial switch is guaranteed to be correct, as it is given by the first measurement data point by definition. The results of the backward and forward iteration will occur chronologically alternating, see Fig. 6.10. The time between the switch detection in backward direction and the switch detection in forward direction is guaranteed to frame the true switching time.

The true switching time is thus overapproximated as well in forward direction $\hat{k}_{\mathcal{T},j,f}$ as in backward direction $\hat{k}_{\mathcal{T},j,b}$, leading to a switch segment

$$k_{\mathcal{T},j} \in [\hat{k}_{\mathcal{T},j,b}, \hat{k}_{\mathcal{T},j,f}]. \tag{6.67}$$

The measurement data between the switch segments, i.e. for the time steps

$$k \in [\hat{k}_{\mathcal{T},j,f}, \hat{k}_{\mathcal{T},j+1,b}] \tag{6.68}$$

belong to a so called trust segment. The trust segment is guaranteed to contain only measurement data generated by a single subsystem. Verification results calculated based on trust segments are not disturbed by measurement data generated by other dynamic systems. The results of the algorithm converge to the true values for a suitable setting of the enclosed noise ϵ and the enclosing interval width δ.

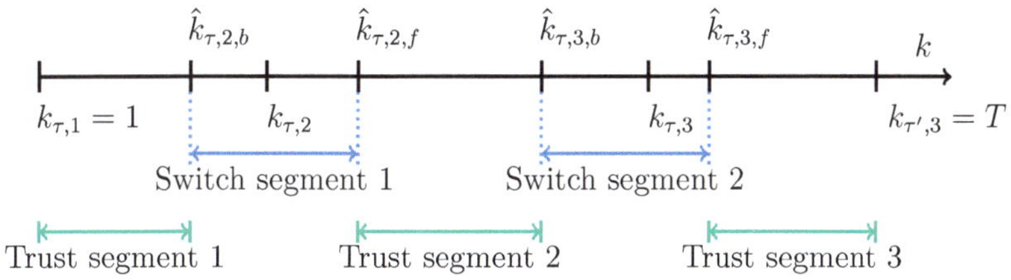

Figure 6.10: Example of the alternating occurrence of trust and switch segments

6.4 Conclusion

The concept of verification of dynamic systems based on Kaucher interval arithmetic introduced in Chapter 4 was extended to hybrid systems in this chapter. Therefore a hybrid model consisting of a discrete time, value continuous, dynamic system and a discrete event state machine were formally introduced. The behavior of the dynamic part is given by parameters that are determined by the discrete event states. As long as there is only one discrete state active, the dynamic verification can be done as in the nonhybrid case. If the active state changes, the procedure has to be adapted. Each segment of measurement data generated by a single active subsystem can be verified individually.

The problem becomes more complex, if it is necessary to determine the segments and the active subsystems within the segments. Three different approaches to solve this problem were introduced.

In a first step, the state signal was measured and the given states were assumed to be mapped with the specification. In a second step, only the switching times were known and it was necessary to determine the respective active states. Conditions of Prager-Oettli-Distinguishability on the specification and the mappability of the measurement data were introduced that ensure the theoretical possibility of determining this information.

In a third step, there was no information at all about the switches and the active states. Therefore an algorithm was developed that is able to determine both, switches and states, in an iterative identification and segmentation procedure. If the introduced property of Prager-Oettli-Segmentability holds, this procedure can be applied successfully.

The verification of the discrete event part can be done in the same way for all three settings. The trace given by the mapped state signal is used to set up the measured state machine which can be compared easily with the defined state machine. Therefore the set of states and the transition function are compared with the nominal values given in the specification.

This approach was partially developed in [Hen15] and published in [Sch17a]. Also it was applied to the practical example of a battery management system [Lem15] and a hybrid braking system [Glü17] and it was presented in [Sch16][Sch18a][Sch19].

7 Extended Kaucher Based Guaranteed Verification

This chapter introduces several extensions to the Kaucher based verification framework developed in the previous chapters. Therefore the point of view is changed from finding "at least one" consistent parameter to calculating the "largest possible set" of parameters within the united solution set. This is beneficial as the calculation of the exact solution set is an NP-Hard problem as stated in Section 3.1.3. This chapter introduces a method that combines the Kaucher based method with optimization techniques to calculate the maximum inner approximation of the feasible set.
This approximation is done using different shapes given by the combination of the objective function and the constraints. Three different shapes are presented and analyzed. The basic approach uses orthogonal enclosure leading to classic interval type borders.
In a second step, this approach is extended to zonotopic sets to exploit specific properties of the measurement data. It is widely known in the interval community that using zonotopic sets avoid large underapproximations that arise from axis parallel orthogonal enclosure [Jau01][Asa06][Pui06][Alt08][Ing09][Mai16][Roe16][Wan17]. Finally all restrictions on the shape of the resulting set are dropped and a general polytope is accepted as result. Although this is the most general shape and yields the largest inner enclosures, it is impossible to describe the result in terms of interval values, thus limiting the usability of this approach.
The last section of the chapter introduces an alternative application of the Kaucher based verification method in an online setting. Therefore the verification procedure is extended to allow an iteratively increasing measurement set. It is thus necessary to provide repeated approximations of the solution set as well as repeated verdicts to evaluate the consistency. This resembles the situation of an ongoing diagnosis process of a running application.

7.1 Solution Set Approximations

Instead of solving the problem whether a particular (nominal) parameter vector is part of the united solution set, the question is now to find the (whole) feasible set given by the measurement data. Therefore the constraints $c_{\mathcal{M}}$ of the feasibility problem given by the measurement data - as defined in Def. 5.6 - are regarded.[13] The feasible set is based only on the measurement data and thus no information about the consistency of the VO is included. The feasible set is defined as:

Definition 7.1 (Feasible Set)
The feasible set $\mathcal{F}$ is given by

$$\mathcal{F} = \left\{ \Theta \in \mathbb{R}^{n_a^*+n_c^*} : c_{\mathcal{M}}^{(i)}(\Theta) \leq 0,\ \forall i \in \{1, 2, \ldots, 2(T - \max(n_a^*, n_c^*))\} \right\}, \tag{7.1}$$

i.e. the set of all parameters Θ that fulfill the constraints given by the measurement data as defined in Def. 5.6.

Note that the genuine feasible set here is given according to Def. 5.6 which represents the united solution set $\sum_{\exists\exists}$. The feasible set is thus not necessarily connected and not necessarily constrained by borders parallel to the axis (see Section 3.1.2).

The feasible set can also be defined by a set of vertexes:

Definition 7.2 (Vertex Based Feasible Set)
The feasible set $\mathcal{F}$ is constrained by the convex hull (see [Bro08, p. 663]), given by the set of vertexes $\mathcal{V} = \left\{V_0, V_1, \ldots, V_{|\mathcal{V}|-1}\right\}$:

$$\mathcal{F}(\mathcal{V}) = \left\{ \Theta \in \mathbb{R}^{n_a^*+n_c^*} : \Theta = \sum_{i=0}^{|\mathcal{V}|-1} \alpha_i V_i \,\middle|\, (\forall i : \alpha_i \geq 0) \wedge \sum_{i=0}^{|\mathcal{V}|-1} \alpha_i = 1 \right\}. \tag{7.2}$$

With this definition it is possible to transfer the setting into an optimization problem based on the vertexes of the convex hull. The shape of the resulting approximation of the feasible set $\mathcal{F}$ becomes a design parameter that is reflected by the objective function and the constraints. The solution of the optimization is thus not necessarily of the same shape as the genuine feasible set.
The approximation of the feasible set can be done in different ways. This thesis uses three different approaches: hyperrectangular approximation, zonotopic approximation and polytopic approximation. All three approaches are introduced in the next sections.

[13] The constraints of the nominal set $c_{\mathcal{N}}$ defined in Def. 5.5 are not necessary for the calculation of the feasible set.

7.1.1 Hyperrectangular Solution Set Approximation

In the hyperrectangular case, the objective function is set up such that the optimization yields the largest hyperrectangle area within the united solution set. This shape represents an interval type result, constrained parallel to the coordinate axis.
The solution set is determined using an optimization setting as given in Prop. 7.1.

Proposition 7.1 (Optimization Based Hyperrectangular Solution Set)
The hyperrectangular approximation of the feasible set $\mathcal{F}^{\square}$ is given by the set of vertexes $\mathcal{V}^{\square}$. The set of vertexes $\mathcal{V}^{\square}$ is calculated based on the measurement data $[\mathbf{U}_{meas}, \mathbf{Y}_{meas}]$ given for the discrete sampling points $k \in \{1, 2, \ldots, T\}$ of a VO. It is defined as the solution of the optimization problem

$$\mathcal{V}^{\square} = \underset{\boldsymbol{\Theta}}{argmax}\left(J^{\square}\left(\boldsymbol{\Theta}\right)\right) \tag{7.3}$$

with the objective function

$$J^{\square}\left(\boldsymbol{\Theta}\right) = \prod_{i=1}^{n_a^*+n_c^*} \left(\overline{\theta}^{(i)} - \underline{\theta}^{(i)}\right) \tag{7.4}$$

that is subject to the constraints

$$c_{\mathcal{P}}^{(i)}\left(\boldsymbol{\Theta}\right) := \underline{\theta}^{(i)} - \overline{\theta}^{(i)} \leq 0 \tag{7.5}$$

$$c_{\mathcal{M}}(\boldsymbol{\Theta}) \leq 0 \tag{7.6}$$

for $i = \{1, 2, \ldots, n_a^* + n_c^*\}$. Thereby $c_{\mathcal{P}}$ constrains the solution to be proper and $c_{\mathcal{M}}$ is given by the measurement data according to Def. 5.6.

Proof:
The result of the optimization problem (7.3) is given by the set of vertexes $\mathcal{V}^{\sqcup}$ that defines the hyperrectangular approximation $\mathcal{F}^{\square}$. The hyperrectangular approximation is proper in all dimensions due to (7.5). The united solution set is given by the measurement data and consists of all parameters that fulfill (7.6) according to Def. 5.6. As the optimization problem (7.3) is constrained by (7.6), all elements of the resulting set of vertexes are part of the united solution set $\sum_{\exists\exists}$ and thus

$$\mathcal{F}^{\square} = \mathcal{F}\left(\mathcal{V}^{\square}\right) \subseteq \mathcal{F} = \textstyle\sum_{\exists\exists} . \tag{7.7}$$

□

In general, an n-dimensional parameter space leads to $|\mathcal{V}^{\square}| = 2^n$ different vertexes that determine the solution set. In the hyperrectangular case, the set of vertexes $\mathcal{V}^{\square}$ is directly given by the interval type parameter vector as illustrated in Example 7.1.

Example 7.1:

For an $n = 2$ dimensional interval type parameter vector

$$\boldsymbol{\Theta} = \left[\boldsymbol{\theta}^{(1)},\ \boldsymbol{\theta}^{(2)}\right]^T \tag{7.8}$$

the resulting rectangle is given by the set of $|\mathcal{V}^{\square}| = 2^n = 4$ vertexes:

$$\mathcal{V}^{\square} = \left\{\left[\underline{\theta}^{(1)},\ \underline{\theta}^{(2)}\right],\left[\overline{\theta}^{(1)},\ \underline{\theta}^{(2)}\right],\left[\overline{\theta}^{(1)},\ \overline{\theta}^{(2)}\right],\left[\underline{\theta}^{(1)},\ \overline{\theta}^{(2)}\right]\right\}. \tag{7.9}$$

Without the constraint set $c_{\mathcal{P}}$ given in (7.5), the setting can lead to improper solutions. These solutions cannot be interpreted and therefore are useless with respect to a real system. This is due to the fact that there is no width defined for improper intervals. Nevertheless, the objective function is defined on infimum and supremum of the intervals and can thus be evaluated even for improper intervals. In case of an even number of parameters being improper, the area multiplication in (7.4) yields a positive value. This value can be increased to infinity in the "improper direction" leading to impossible results. Such improper solutions are not suitable for the verification setting, as the measurement was obtained from a real system, with real generating parameters. An illustration of the rectangular inner enclosure is given in Example 7.2.

Example 7.2:

Assume the united solution of the measurement data to be given as the blue shape in Fig. 7.1. The solution is in general not unique because there might be other possible solutions of the same size. Two possible area maximal inner enclosures are given by the green rectangles. Note that this example is showing the basic concept of the enclosure and thus the denoted values are chosen arbitrary.

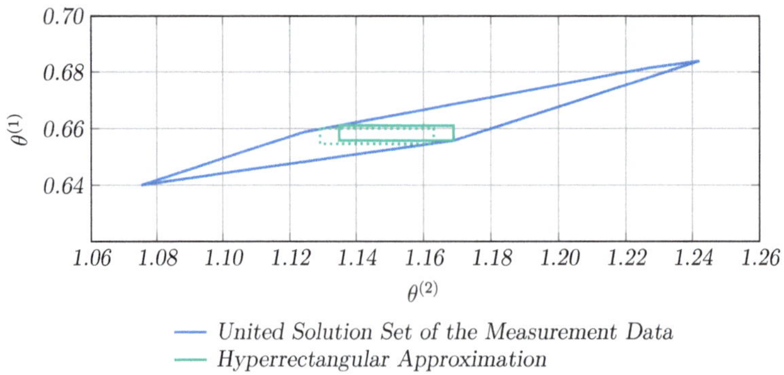

Figure 7.1: Area maximal axis parallel hyperrectangular inner approximation of the united solution set

7.1.2 Zonotopic Solution Set Approximation

In this approach, the united solution set is constrained by a set of hyperstripes in the parameter space, generated by the measurement data. The orientation of these hyperstripes is similar for consistent measurement data. This leads to a united solution set showing a zonotopic shape which is depicted as the shaded area in Fig. 7.2.

The formal description of the hyperstripes is done based on [Ble11] such that each line of the regressor matrix defines a set of feasible parameters $\mathcal{F}^{(k)}$. Thereby $\mathcal{F}^{(k)}$ denotes the feasible set at the specific time step k. The feasible hyperstripes are given by

$$\mathcal{F}^{(k)} = \{\Theta \in \mathbb{R}^{n_a^*+n_c^*} : -\delta^a \leq B_{meas,k} - A_{meas,k}\Theta \leq \delta^a\} \tag{7.10}$$

with δ^a being the width of the interval enclosure, i.e. the maximum absolute sensor fault. The hyperstripe can be written in normalized form as

$$\mathcal{F}^{(k)} = \left\{\Theta \in \mathbb{R}^{n_a^*+n_c^*} : \left|\frac{B_{meas,k}}{\delta^a} - \frac{A_{meas,k}}{\delta^a}\Theta\right| \leq 1\right\}. \tag{7.11}$$

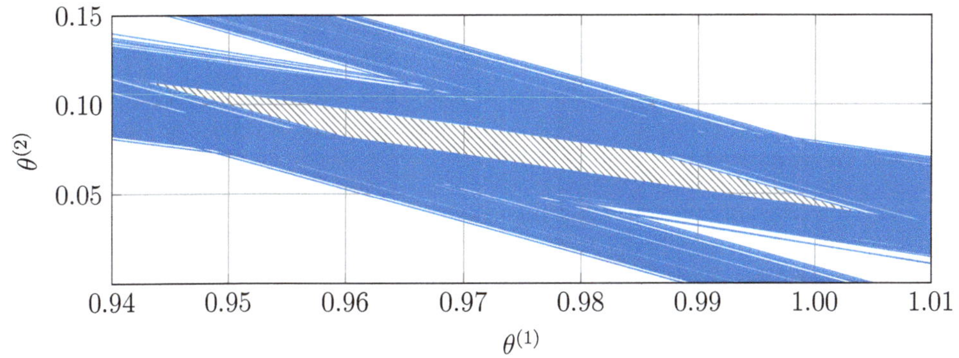

Figure 7.2: Constraints showing the general zonotopic shape (shaded area) of the united solution set in the case of two parameters

The feasible set $\mathcal{F}_k$ that includes all measurement information up to time step k, can be determined recursively by intersecting the hyperstripes $\mathcal{F}^{(k)}$:

$$\mathcal{F}_k = \mathcal{F}_{k-1} \cap \mathcal{F}^{(k)}. \tag{7.12}$$

Therefore the resulting feasible set for all available measurement data is given by $\mathcal{F}_T$. The approximation of this set can be done by a zonotopic set with a very good fit. Definition 7.3 describes such a general zonotopic set $\mathcal{Z}$ as given in [Ble11].

Definition 7.3 (Zonotopic Set)

A zonotopic set $\mathcal{Z}$ is constrained by the convex hull of the set of vertexes $\mathcal{V}^\diamond = \{V_0, V_1, \ldots, V_{|\mathcal{V}^\diamond|-1}\}$. The calculation of the vertexes is done by

$$\mathcal{V}^\diamond = P^0 \oplus H^0 \boldsymbol{K}^V = \{P^0 + H^0 \boldsymbol{z} : \boldsymbol{z} \in \boldsymbol{K}^V\} \tag{7.13}$$

with the center of the zonotope $P^0 \in \mathbb{R}^{(n_a^* + n_c^* \times 1)}$, the radius matrix $H^0 \in \mathbb{R}^{(n_a^* + n_c^* \times V)}$ and a unitary box $\boldsymbol{K}^V$ composed of an arbitrary number of V unitary interval vectors $\boldsymbol{K} = [-1,\ 1]$.

The $\oplus$-operator denotes the Minkowski-Sum [Ber08, p. 291] that is used to calculate the vertexes of the zonotope. This means that the center P^0 is added to the given unitary vectors (i.e. corners) that are scaled by the respective radius value. The adaption to the zonotopic approximation of the feasible set is done in Def. 7.2, leading to the notation $\mathcal{F}^\diamond$. The respective optimization problem is set up such that a zonotopic approximation $\mathcal{F}^\diamond$ of the united solution $\sum_{\exists\exists}$ is achieved.

Proposition 7.2 (Optimization Based Zonotopic Solution Set)

The zonotopic approximation of the feasible set $\mathcal{F}^\diamond$ is given by the set of vertexes $\mathcal{V}^\diamond$. The set of vertexes $\mathcal{V}^\diamond$ is calculated based on the measurement data $[\mathbf{U}_{meas},\ \mathbf{Y}_{meas}]$ given for the discrete sampling points $k \in \{1, 2, \ldots, T\}$ of a VO. It can be computed based on the optimal scaling parameter

$$\alpha^\diamond = \underset{\alpha}{argmax}\left(J^\diamond(\alpha)\right), \tag{7.14}$$

the initial center P^0 and the radius matrix H^0 given by the outer enclosure of the measurement data. The vertexes of the zonotope $\mathcal{V}^\diamond$ are calculated by

$$\mathcal{V}^\diamond = P^0 \oplus \alpha^\diamond H^0 \boldsymbol{K}^V = \{P^0 + \alpha^\diamond H^0 \boldsymbol{z} : \boldsymbol{z} \in \boldsymbol{K}^V\}. \tag{7.15}$$

The optimization problem consists of the objective function

$$J^\diamond(\alpha) = \alpha \tag{7.16}$$

that is subject to the constraints

$$c_{\mathcal{M}}(\mathcal{V}^\diamond) \leq 0 \tag{7.17}$$

given by the measurement data according to Def. 5.6.

Proof:
The result of the optimization problem (7.14) is given by the set of vertexes $\mathcal{V}^{\diamond}$ that defines the zonotopic approximation $\mathcal{F}^{\diamond}$. The zonotopic approximation is proper in all dimensions as it is based on an initial outer approximation calculated using classic interval arithmetic and subsequently scaled using a point real parameter. The united solution set is given by the measurement data and consists of all parameters that fulfill (7.17) according to Def. 5.6. As the optimization problem (7.14) is constrained by (7.17), all elements of the resulting set of vertexes are part of the united solution set $\sum_{\exists\exists}$ and thus

$$\mathcal{F}^{\diamond} = \mathcal{F}\left(\mathcal{V}^{\diamond}\right) \subseteq \mathcal{F} = \sum\nolimits_{\exists\exists}. \tag{7.18}$$

□

This method has its roots in [Ble11] and [Sch17c] and was developed in [Sch18b]. The initial parameters of the optimization are given by the center P^0 and the radius matrix H^0 which are determined by calculating the outer enclosure as in [Ble11], given in (7.10)-(7.12) extended to Kaucher arithmetic notation.
The initial zonotope to calculate this outer enclosure needs to be chosen suitably. One possibility is to use the nominal region $\mathbf{\Theta}^* = [\Theta_c^* - \Theta_\Delta^*,\ \Theta_c^* + \Theta_\Delta^*]$ expressed as $P^0 = \Theta_c^*$ and $H^0 = I\Theta_\Delta^*$. It is also possible to calculate the point real central solution Θ_c using the center matrizes $A_{meas,c}$ and $B_{meas,c}$. The initial zonotope is then given by $P^0 = \Theta_c$ and $H^0 = I\epsilon$ with an arbitrary small value $\epsilon > 0$.

Each measurement interval is iteratively interpreted as a hyperstripe containing the possible parameters. The intersection between the hyperstripe and the zonotope is calculated and the common region is used to calculate the new radius matrix. This procedure leads to a zonotopic outer enclosure of the feasible parameter set. Due to the repeated calculation of outer enclosures, the area of the zonotope might grow with the considered measurement data.
Starting from the final outer enclosure that frames the intersection of all available measurement data, the scaling factor α is minimized until all vertexes of the zonotope are part of the united solution. This can be checked using Prop. 4.1.
The center point of the zonotope is not moved by the shrinking procedure. The idea of maximum possible parameter variability within the zonotope is realized by the objective function (7.16) that maximizes the scaling factor α.
Additional assumptions on the optimization problem are:

- The interval solution has to be bounded to one orthant.
- All values of the input signal need to have the same sign, either all positive or all negative.

The first assumption is due to the fact that intervals containing zero can be interpreted erroneously as inverse elements and thus cancel the influence of some parameters. The second assumption is necessary to prevent increasing intervals that may arise even when Kaucher interval arithmetic is used.
A sketch of a zonotopic inner approximation is given in Example 7.3.

Example 7.3:

Assume the united solution of the measurement data to be given as the blue shape in Fig. 7.3. The center of the zonotope is depicted by the green circle. It is given by the center of the outer enclosure that is used as initial zonotope. The algorithm does not change the center, leading to the area maximal inner enclosure given by the green zonotope.
Note that there might be solutions of equal or larger size possible for a different zonotope center. However the choice of an optimal center is not included in the current version of the algorithm. The denoted values are chosen arbitrary in this example.

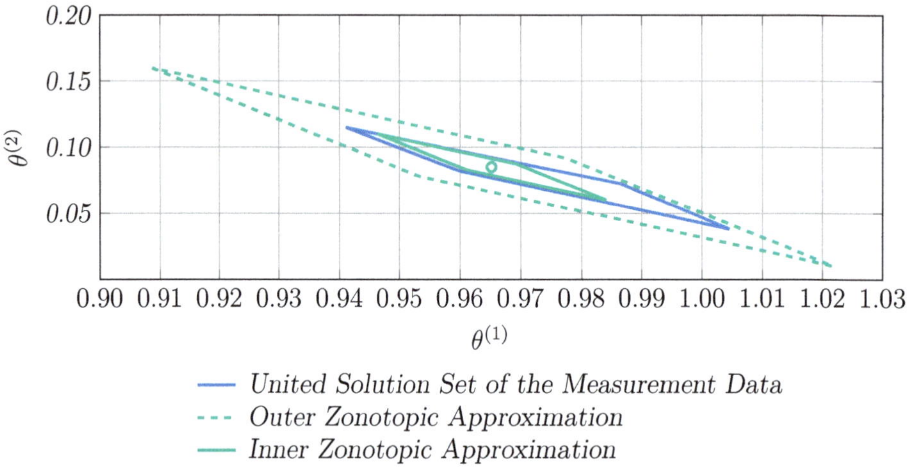

Figure 7.3: Exemplary area maximal zonotopic approximation of the united solution set

7.1.3 Polytopic Solution Set Approximation

The most general enclosure is given by the shape of a convex polytope. However in this case the solution is represented by a list of points instead of a mathematical description such as intervals (in the hyperrectangular case) or center and radius matrix (in the zonotopic case). The only conditions on a valid polytopic enclosure consists of the requirement that points of the list are part of the united solution set $\sum_{\exists\exists}$ according to Prop. 4.1 and that the resulting polytope is convex. The computational effort of the polytopic approximation is in the same order of magnitude as the computational effort of the hyperrectangular approximation. The respective solution of the optimization problem is given in proposition Prop. 7.3.

Proposition 7.3 (Optimization Based Polytopic Solution Set)
The polytopic approximation of the feasible set $\mathcal{F}^{\pentagon}$ is given by the set of vertexes $\mathcal{V}^{\pentagon}$. The set of vertexes $\mathcal{V}^{\pentagon}$ is calculated based on the measurement data $[\mathbf{U}_{meas}, \mathbf{Y}_{meas}]$ given for the discrete sampling points $k \in \{1, 2, \ldots, T\}$ of a VO.
It is defined as the solution of the optimization problem

$$\mathcal{V}^{\pentagon} = \underset{\Theta^{\pentagon}}{argmax}\left(J^{\pentagon}\left(\Theta^{\pentagon}\right)\right) \tag{7.19}$$

with the objective function

$$J^{\pentagon}\left(\Theta^{\pentagon}\right) = area\left(\Theta^{\pentagon}\right). \tag{7.20}$$

The function area $(\cdot)$ calculates the hypervolume of a polytope given by a list of points $\Theta^{\pentagon}$. All points of the list, i.e. vertexes of the set $\mathcal{F}^{\pentagon}$, are subject to the constraints

$$c_{\mathcal{M}}(\Theta^{\pentagon}) \leq 0 \tag{7.21}$$

with $c_{\mathcal{M}}$ given by the measurement data according to Def. 5.6.

Proof:
The result of the optimization problem (7.19) is given by the set of vertexes $\mathcal{V}^{\pentagon}$ that defines the polytopic approximation $\mathcal{F}^{\pentagon}$. The polytopic approximation is proper in all dimensions as it is the convex hull of $\mathcal{V}^{\pentagon}$. The united solution set is given by the measurement data and consists of all parameters that fulfill (7.21) according to Def. 5.6. As the optimization problem (7.19) is constrained by (7.21), all elements of the resulting set of vertexes are part of the united solution set $\sum_{\exists\exists}$ and thus

$$\mathcal{F}^{\pentagon} = \mathcal{F}\left(\mathcal{V}^{\pentagon}\right) \subseteq \mathcal{F} = \sum\nolimits_{\exists\exists}. \tag{7.22}$$

□

The choice of the vertexes of the initial list is of minor importance. Possible choices are the vertexes of the nominal set, the vertexes of a zonotopic outer enclosure or the vertexes given by the central solution disturbed by a small parameter $\epsilon > 0$.

Example 7.4:

Assume the united solution of the measurement data to be given as the blue shape in Fig. 7.4. A possible inner enclosure is given by the green polytope.
Note that this solution is not unique. There might be other possible solutions of the same size. Again, this example is showing the basic concept of the enclosure and thus the denoted values are chosen arbitrary.

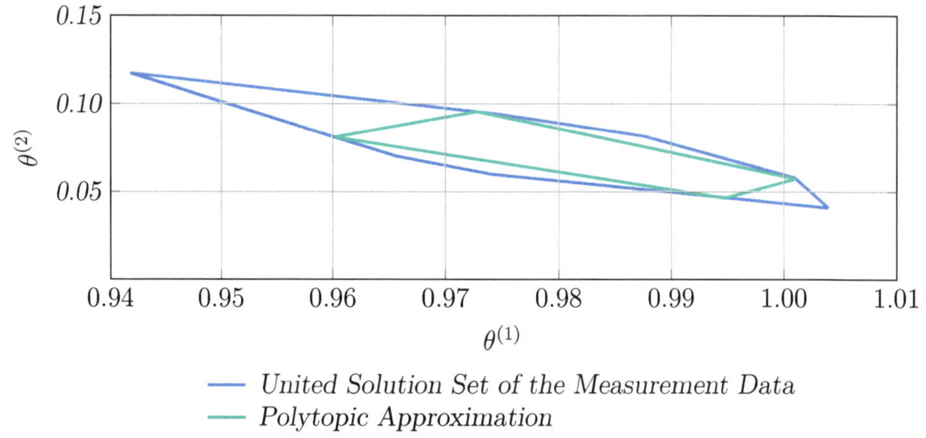

Figure 7.4: Exemplary polytopic inner approximation of the united solution set

7.2 Kaucher Based Diagnosis

The consistency for interval type systems was introduced in Chapter 5. Prop. 5.4 focuses on basic consistency and includes the main result that forms the foundation of the considerations in this chapter. All other assumptions and definitions are assumed to be fulfilled and applicable as well.
In case the calculation of the verification method can be fast enough with respect to the regarded dynamic system, the method can be applied in an online setting to tackle the diagnosis problem. In the resulting diagnosis setting, the interpretation of the problem includes the temporary feasible set $\mathcal{F}_k$ that is approximated using one of the three approximation shapes introduced in Section 7.1. The intersection between the temporary feasible set $\mathcal{F}_k$ and the nominal set $\mathcal{N}$ forms an inner enclosure of the consistent set for the whole measurement time according to Sec. 5.2.

Proposition 7.4 (Approximation Based Basic Consistency)
The input-output behavior given by all available interval type enclosures of T measurement values

$$[\mathbf{U}_{meas}, \mathbf{Y}_{meas}] = \left[\langle \mathbf{u}_{meas,k} \rangle_{k=1}^{T}, \langle \mathbf{y}_{meas,k} \rangle_{k=1}^{T} \right] \tag{7.23}$$

is basic consistent with the nominal system specified by an interval type specification

$$S_i^* = \{ \mathbf{\Theta}^*, n_a^*, n_c^*, U_{init}^*, Y_{init}^* \}, \tag{7.24}$$

if the intersection between nominal set $\mathcal{N}$ and temporary feasible set $\mathcal{F}_k$ is nonempty, i.e.

$$\mathcal{N} \cap \mathcal{F}_k \neq \emptyset. \tag{7.25}$$

Proof:
The nominal set $\mathcal{N}$ consists of all parameters within the specification $\mathbf{\Theta}^*$. Basic consistency according to Prop. 5.4 is given, if there is at least one parameter $\Theta \in \mathbf{\Theta}^*$ within the specified parameter set that is able to explain the measurement data.
According to Def. 7.1, all parameters of the feasible set $\mathcal{F}_k$ are able to explain the measurement data for $k \in \{1, 2, \ldots, T\}$.
The intersection between the nominal set $\mathcal{N}$ and the feasible set $\mathcal{F}_k$ contains parameters that are both, part of the nominal set and part of the feasible set. If this intersection is nonempty, there is at least one parameter that is consistent according to Prop. 4.1. □

It is in general not possible that the method yields full consistency in the diagnosis setting. However, the approximation methods described in the previous section can be used to verify dynamic systems given by an interval type specification S_i^*.[14]
In the diagnosis setting, an approximation of the feasible set is used to calculate an inner enclosure of the consistent set. The verification result is guaranteed in the sense of this thesis as it is based on Kaucher arithmetic. The definition of consistency uses the united solution set as given in Def. 3.1. The choice of vertex points used during the verification is done based on the introduced optimization procedure hence ensuring an effective coverage of the feasible area given by the measurement data. The shape of the calculated solution is determined directly by the setup of the objective function and the additional constraints. Furthermore, the calculated inner approximation aims at finding an area maximal representation of the united solution set in terms of the defined approximation shape. Thus the feasible set represents the maximum possible parameter variability in the given setting that is guaranteed to be able to explain the measurement data.

[14] This method is also applicable to a point real specification. Nevertheless a point real specification can be verified directly by Prop. 4.1 and does not need the optimization based extensions in order to approximate the feasible set.

It is assumed that the diagnosis algorithm is running in parallel with the VO. The diagnosis system is supplied with a new measurement value in each sampling cycle which is used to calculate a temporary result. This leads to the diagnosis algorithm as given in Fig. 7.5.

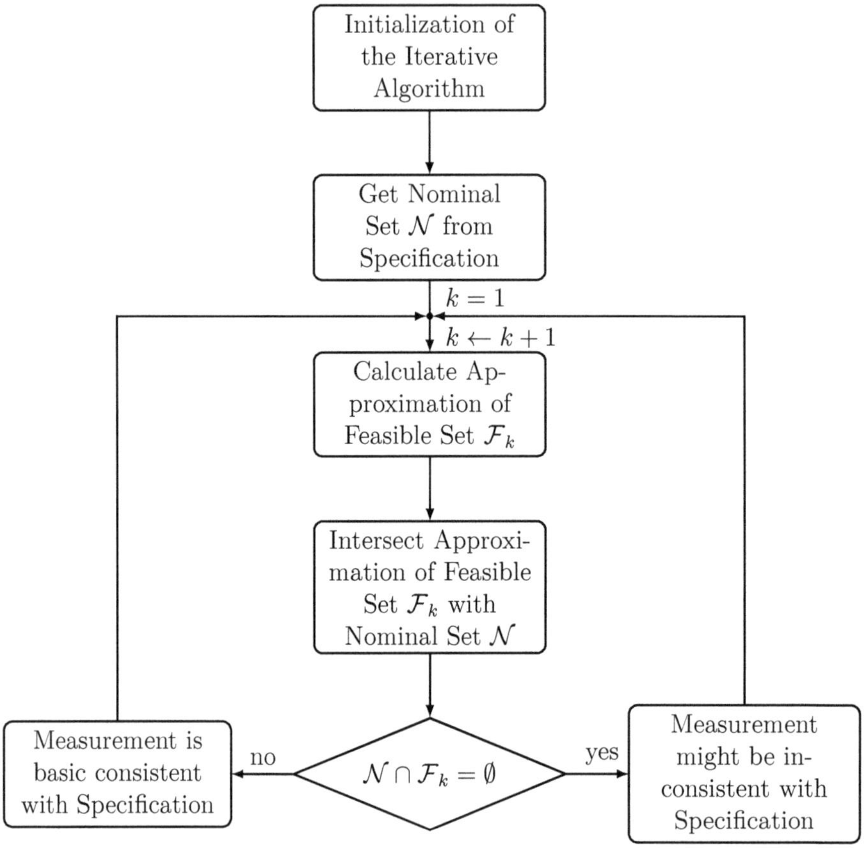

Figure 7.5: Diagnosis algorithm

A VO that is basic consistent for all measurement data at the final time T, considering all measurement data with $k < T$ according to Prop. 7.4, is assumed to be also basic consistent for $\tilde{T} < T$.
If the algorithm yields basic consistency between the measurement data and the specification, the verdict is guaranteed to be correct. Therefore there are no hidden faults i.e. no type II errors possible.
If the algorithm yields inconsistency, the verdict might be erroneous, i.e. showing a false alarm. This is due to the used inner approximation of the feasible set which does not necessarily cover all feasible parameters.

Three exemplary settings and the resulting solution sets are depicted in Fig. 7.6. The black lines depict the constraints given by the measurement data that frame the united solution. The nominal parameters are given by the black dashed rectangle.

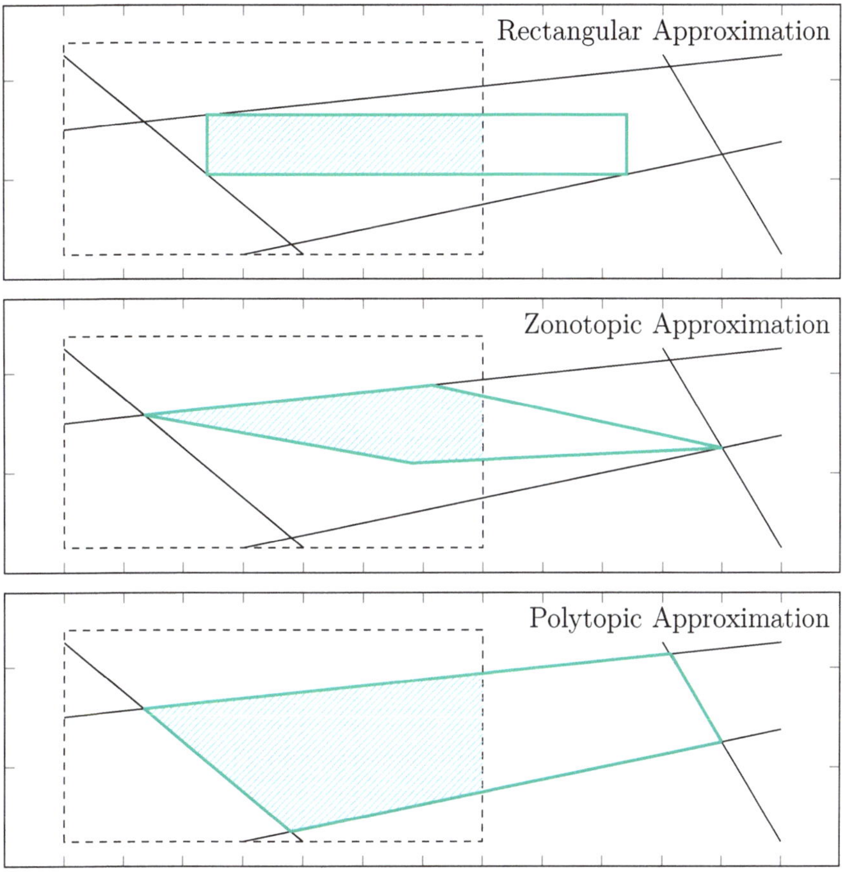

Figure 7.6: Exemplary results for the three different approximation shapes

The first plot shows an inner approximation using an rectangular shape. It can be seen that the location and shape of the rectangle is not unique and that other rectangles of the same size might be possible within the united solution. The second picture shows the approximation by a zonotope. The general fit of the zonotope is better as it can follow the contour of the constraints to some extent. The third figure shows the approximation by a polytope. This setting shows the best fit, even though the resulting shape is given by a list of four points instead of an algebraic description. All three examples show an intersection between the nominal set and the approximation of the united solution set, given by the shaded areas. The regarded example setting thus shows a situation with guaranteed basic consistency between measurement and specification.

7.2.1 The Center Misplacement Effect

The zonotopic approximation leads to the best tradeoff between accuracy and ease of description. This shape is chosen for the further explanations of the diagnosis approach. As already mentioned, false alarms are possible for all regarded shapes due to the usage of inner approximation of the feasible set.
In the special case of a zonotopic approximation, false alarms can also result from the so called Center MisPlacement Effect (CMP). CMP denotes the effect of disregarding feasible results due to a bad center and shape of the zonotope.
An exemplary situation showing a CMP effect is given in Fig. 7.7.

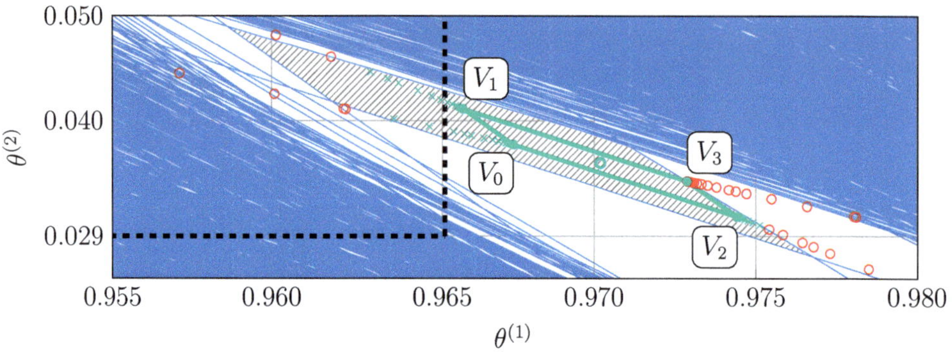

Figure 7.7: Center misplacement effect

The blue lines depict the constraints given by the measurement data. The center of the zonotope is given by the green circle within the green zonotope. Center and shape of the zonotope were calculated from an outer approximation of the solution set as in [Ble11]. Afterwards the scaling parameter α was adapted such that all vertexes of the zonotope are within the united solution of the problem, given by the shaded area. Temporary feasible vertexes of the zonotope that were checked during this optimization procedure are given by green crosses.[15]
Due to the location of the center and the shape of the zonotope, the scaling factor α needs to be very small to ensure that the vertexes V_2 and V_3 stay within the united solution set.
The resulting zonotope does not intersect with the nominal set, from which the lower right corner is depicted by the dashed black square in the left of the figure.
Even though the final vertex is rejected there are consistent intermediate vertexes (green crosses within the nominal set) that are rejected because the respective temporary zonotope was rejected as not all vertexes were part of the united solution for this value of the scaling parameter α.

Certain measures can be taken to avoid CMP. The most straight forward is to move the center of the zonotope or to change its shape. However, in general it is not trivial to chose a suitable change in center and shape algorithmically.

[15] Temporary vertexes of the zonotope that were checked during the optimization procedure and found to be infeasible are marked as red circles.

As the method is also designed for higher dimensions and should work as automatically and autonomously as possible, it is neither possible nor suitable to visualize the setting and request a human operator to adapt the center or the shape.
An algorithmic workaround is given by the use of intermediate vertexes that are checked during the optimization. The optimization can be stopped as soon as there is at least one feasible intermediate vertex detected, i.e. if a green cross is evaluated. Nevertheless this will lead to point-wise results instead of a feasible set of a given shape. An effective way to avoid CMP is provided by the enclosure of the nominal set in the constraints of the optimization problem. However, this will also pose higher restrictions on the feasible set.

7.3 Conclusion

This chapter introduced extensions to the Kaucher based verification method developed in the previous chapters.
The first extension is given by calculating the largest inner approximation of the united solution set. This was done using an optimization setting. The precise setup of the objective function and the constraints determines the resulting shape of the so called feasible set. The feasible set was approximated using three different geometric shapes (hyperrectangle, zonotope, polytope). The resulting set is an inner approximation of the united solution set, meaning that there are several equivalent approximations within the united solution set.

The second extension is given by the application of the verification method in a diagnosis setting. The diagnosis algorithm was introduced in an iterative setting, calculating temporary results for each sampling cycle. Due to the inner approximation used, the results can show a false alarm. A vivid cause of a false alarm is given by the center misplacement effect, that can occur in case of a zonotopic approximation. The applicability of the Kaucher based diagnosis method depends on the specific settings of the regarded system. Therefore it is necessary to evaluate each system individually before applying Kaucher based diagnosis.

8 Application and Results

This chapter presents the application of the developed methods to tank systems with a varying number of tanks and an adjustable set of connections. The settings are analyzed in simulation and practice.

Tank systems form a class of wide spread theoretic control applications. The basic setting is usually given by one or more tanks with a nominal outflow and a controllable inflow. The goal is to adjust or maintain a nominal height of the fluid in a tank. The process can be disturbed by additional leakages or inflows or by congestions in the in- or outflow pipes. Depending on the specific setup, cross flows between tanks are possible. Those cross flows are in general dependent on the current filling level of the concerned tanks. All measurement data is obtained using a real three-tank process available at the Institute of Control Systems (IRS) at the Karlsruhe Institute of Technology (KIT). A picture of the lab setting is shown in Fig. 8.1. The algorithmic calculations are done on a `Lenovo ThinkPad T460s` powered by an `Intel`® `Core`™ `i7-6600` CPU using `12`GB main memory. The implementation was done in `Matlab`© `2012b`.

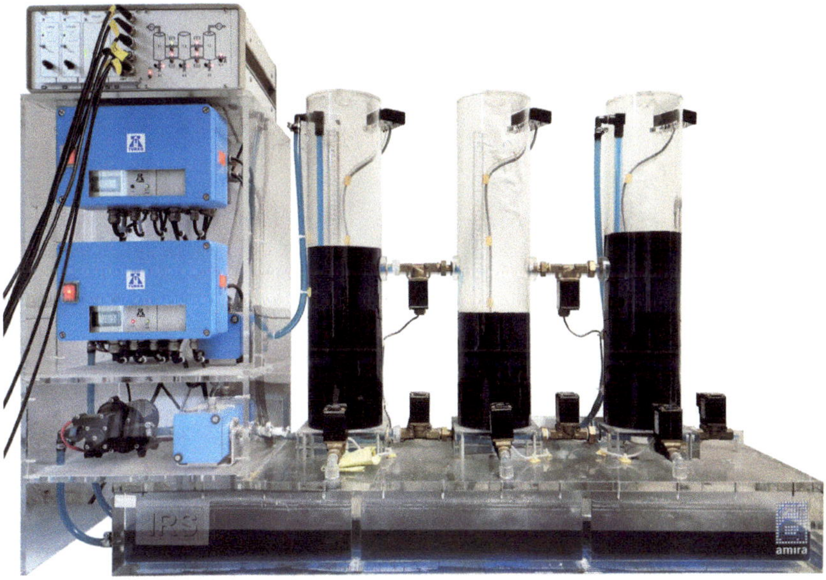

Figure 8.1: Three-tank process lab setting at the Institute of Control Systems (IRS)

The following subsections set up the dynamic models and introduce the properties and possibilities of the different scenarios. First the methods of Chapter 5 "Guaranteed Verification of Interval Type Systems" are applied to real measurement data obtained from a single-tank process.
Second, the application is extended to a two-tank setting showing hybrid behavior. A mapped state signal is used to demonstrate the basic functionality of the method developed in Chapter 6 "Guaranteed Verification of Hybrid Systems". The results are discussed and interpreted.
Third, the diagnosis method developed in Chapter 7 "Extended Kaucher Based Guaranteed Verification" is applied to simulation data of a four-tank process. Three different fault types are used to demonstrate the fault detection properties of the Kaucher based method. Several fault intensities demonstrate the possibility of the method to detect even very small faults.
Finally the diagnosis method is applied to real measurement data provided by a single-tank process. It is shown that all of the regarded faults can be detected successfully using the introduced methods.

8.1 Application: Guaranteed Verification for Interval Type Systems (Single-Tank)

The basic setting is given by a single-tank process which is sketched in Fig. 8.2.

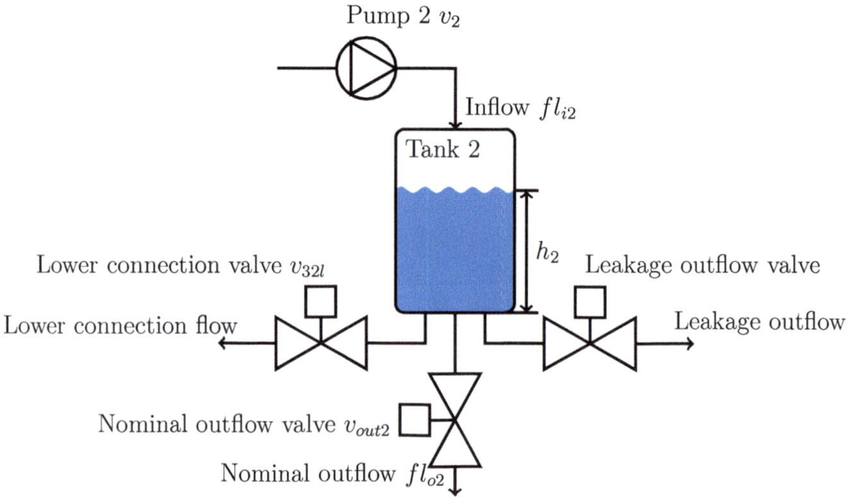

Figure 8.2: Sketch of the single-tank

It is possible to set up a single-tank system consisting only of tank 2 by closing the respective valves in the three-tank lab system.[16] All valves in the system are binary valves which are only open or closed. The height h_2 of tank 2 is measured, as well as the flow v_2 of pump 2.

[16] Note that the number of the tanks in the lab setting is (from left to right) 1 - 3 - 2.

The outflow of each tank is governed by the formula of Torricelli [Tip00, p. 360] which leads to the nonlinear time continuous dynamic of a single-tank

$$\frac{\mathrm{d}h_2(t)}{\mathrm{d}t} = -\frac{1}{A_2}\underbrace{a_2\sqrt{2gh_2(t)}}_{\text{outflow}} + \frac{1}{A_2}\underbrace{\gamma_2 v_2(t)}_{\text{inflow pump 2}}. \tag{8.1}$$

with the outflow pipe cross section a_2, the tank cross section A_2, the gravitational force g and the constant γ_2 according to Appendix G, Tab. G.1. The following simplifications and the resulting model equations are based on the considerations in [Ble11].

The model is discretized using the Euler method with sampling time Δt leading to

$$h_{2,k} = h_{2,k-1} - \frac{a_2}{A_2}\sqrt{2gh_{2,k-1}}\Delta t + \frac{\gamma_2}{A_2}v_{2,k-1}\Delta t + e_{2,k} \tag{8.2}$$

where $e_{2,k}$ is the additive disturbance including sensor and discretization faults. Equation (8.2) is reformulated to the pseudo linear regressor form

$$\varphi_k \theta_k = y_k \tag{8.3}$$

with

$$\varphi_k = h_{2,k-1} \tag{8.4}$$

$$\theta_k = 1 - \frac{a_2}{A_2}\sqrt{\frac{2g}{h_{2,k-1}}}\Delta t + \frac{e_{2,k}}{h_{2,k-1}} \tag{8.5}$$

$$y_k = h_{2,k} - \frac{\gamma_2}{A_2}v_{2,k-1}\Delta t. \tag{8.6}$$

It can be seen that the parameter θ_k given in (8.5) is depending on the height $h_{2,k-1}$ which renders it time variant. The range of the time variant parameter can be interpreted as an interval set that includes all possible parameter values as well as some spurious solutions. The interval enclosure of the parameter is given by the bounding box of the time variant parameter.

It is possible to calculate the interval enclosure of the time variant parameter θ_k for a specific nominal setting ($e_{2,k} = 0$). This setting consists of a given operation range $h_2 \in [h_{2,min}, h_{2,max}]$ and a fixed sampling time Δt. The calculated parameter interval can then be used as the nominal set in further considerations.

Based on the tank properties given in Appendix G, Tab. G.1 it is possible to set up the parameter range for a nominal and a faulty tank configuration. To realize the faulty behavior all available valves are opened. This means the leakage outflow valve, the lower connection valves and the nominal outflow valves. The resulting height dependent parameter θ_k is depicted in Fig. 8.3 and can be used a priori to reason about detectability.

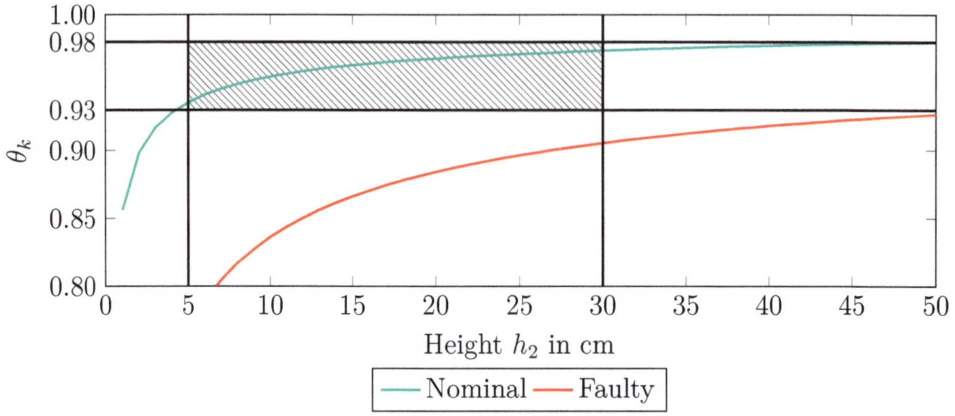

Figure 8.3: Values of θ_k depending on h_2 for different outflow configurations

It can be seen that there is a gap between the nominal behavior (green line) and the faulty behavior using all possible outflows (red line). This leads to the hypothesis that it is possible to separate the nominal behavior from the faulty tank configuration using the method from Chapter 5. Due to the upper valve position $h_{vu} = 30\text{cm}$, the depicted values are valid in the case of a closed upper connection valve v_{23u} only. The nominal parameter $\boldsymbol{\theta}^* = [0.93,\ 0.98]$ is used for tank levels in the range $h_2 \in [5,\ 30]$ cm.

The hypothesis is verified using the following steps: First, a nominal description i.e. a specification, of the system in the necessary ARX form is set up. Second, the system is implemented i.e. the real tank is manufactured and used to collect measurement data. Third, a faulty version of the system is implemented, i.e. unspecified outflows are added to the tank by opening the respective valves.
The collected measurement data for h_2 is then enclosed by interval values

$$\boldsymbol{h}_{2,k} = \left[h_{2,k}(1 - \delta^r_{h_2}),\ h_{2,k}(1 + \delta^r_{h_2})\right] \tag{8.7}$$

and used to verify the correct system with respect to the nominal parameters. Finally, measurement data from the faulty system is used to show that it is not possible to verify the faulty behavior using the nominal parameters.

The initial height is set to $h_{2,min} = 5\text{cm}$, the nominal outflow valve v_{out2} is open and pump 2 is constantly running, providing the maximal inflow of $v_2 = 6.5\text{l/m}$. The resulting nominal setting is depicted in Fig. 8.4. It can be seen that it is possible to verify the measurement of the fault free system supposing a relative fault of $\delta^r_{h_2} = 0.02$ for the measurement signal of the height h_2. The basic consistency as introduced in Chapter 5 is used to calculate the results. The calculation time for the algorithm is $T_{calc} = 16.8\text{s}$ which is less than the duration of the experiment $T = 60\text{s}$.

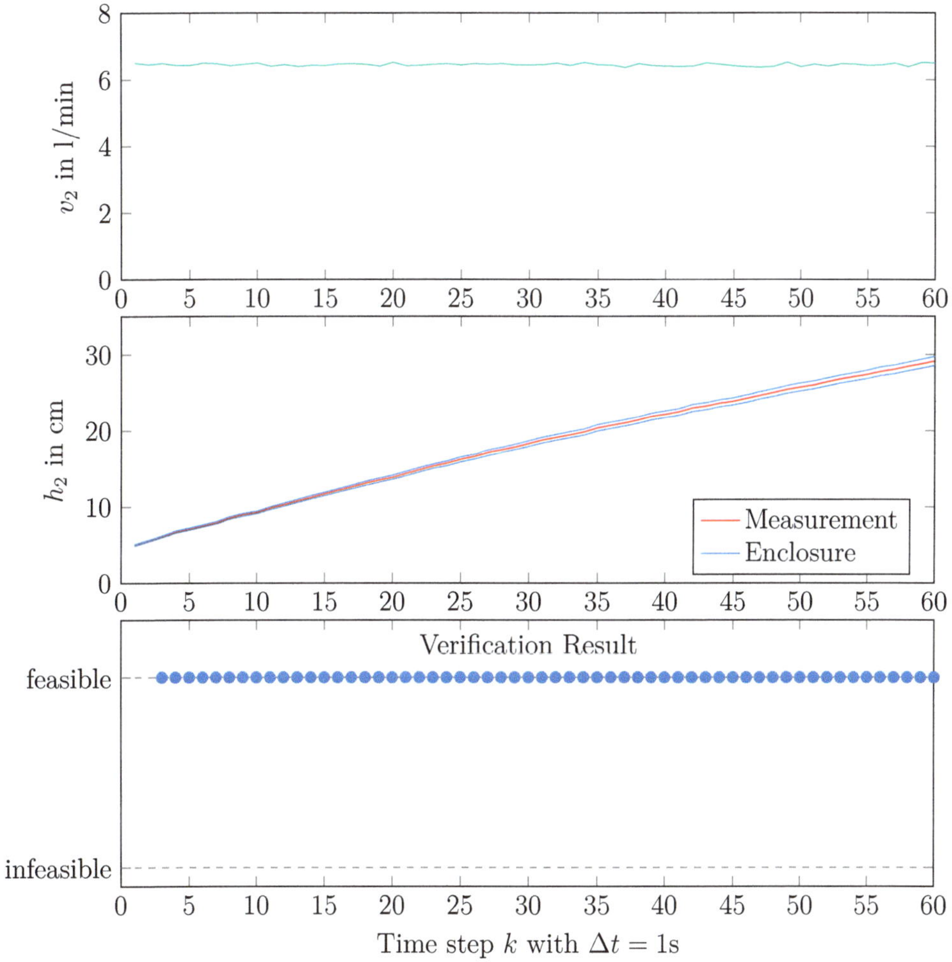

Figure 8.4: Implementation of the single-tank setting that is basic consistent with the specification

Now a faulty implementation of the tank system is considered. Therefore the faulty setting is realized by opening the leakage outflow, the lower connection and the nominal outflow valves which leads to a major change in the system dynamics.
The changed dynamics are not able to reach the nominal final height of $h_{2,60} = 29.1\text{cm}$ as the maximum pump flow cannot compensate the additional outflow. The resulting measurement data is depicted in Fig. 8.5. It is again analyzed using a relative fault of $\delta^r_{h_2} = 0.02$. Fig. 8.5 shows that it is not possible to verify the faulty implementation for $k \geq 4$. The necessary calculation time is $T_{calc} = 47.5\text{s}$.

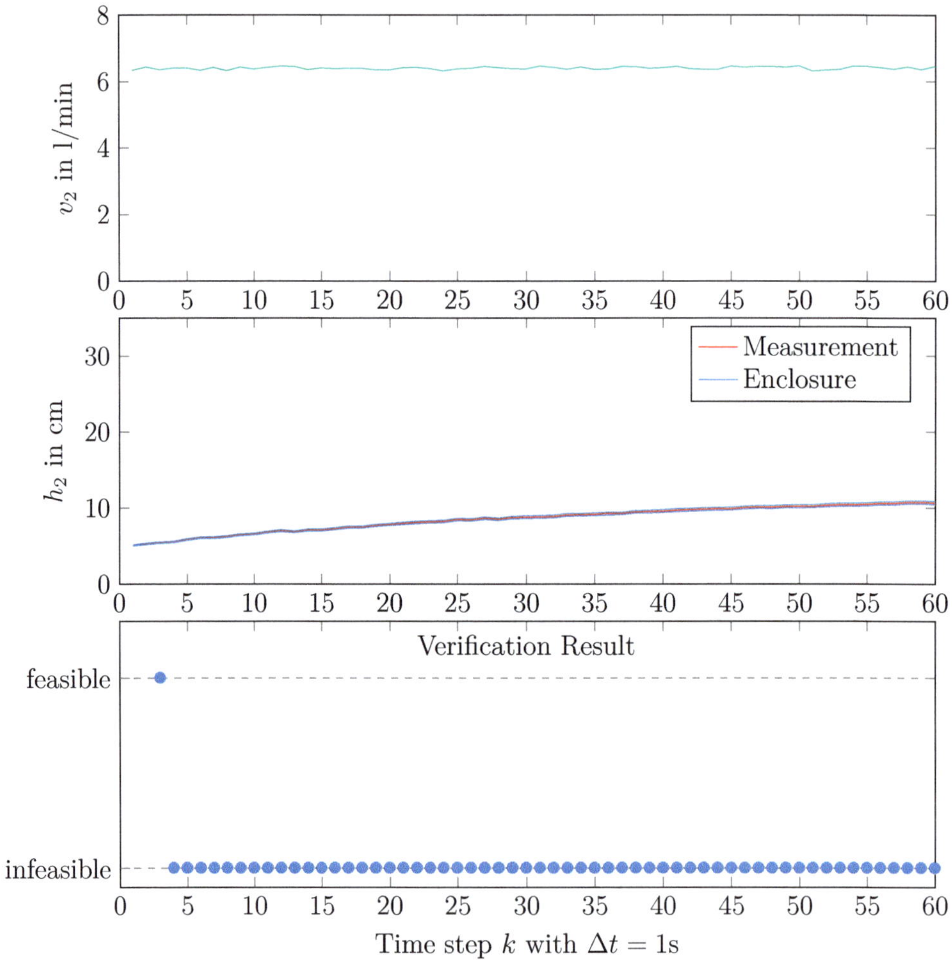

Figure 8.5: Implementation of the single-tank setting that is inconsistent with the specification due to additional outflows

The change in the system dynamics has to be strong to prevent the verification of the faulty system. If there is only a slight change in the system dynamics, the faulty system is verified because the new behavior can be explained with the range of the nominal parameter set. This is the case, if a faulty system is implemented consisting of other combinations of outflows and connection valves except the introduced setting.
When having a closer look on the tank parameters, this is rather intuitive. The connection valves cross section a_{32l} and a_{32u} are the same as the nominal outflow a_2. The leakage a_{leak} is only slightly bigger that the nominal outflow. Thus each individual valve leads to none, respectively very slight variations with respect to the nominal behavior. Any combination of two outflows is also verified using the original setting.

This concludes the first application example concerning the guaranteed verification for interval type systems as introduced in Chapter 5. It is clear that the results are depending on the value of the used relative fault $\delta^r_{h_2}$. Increasing the fault increases the variety of enclosed trajectories and can thus be interpreted as increasing the available system behavior. This leads to a higher chance to achieve nominal behavior enclosed in the measurement data and thus to verify the system.

8.2 Application: Guaranteed Verification for Hybrid Systems (Two-Tank)

The single-tank-system can be extended by adding another tank, connected via two horizontal valves. There is no additional pump and thus no external inflow to the new tank. However the new tank has a nominal outflow and it is possible to open and close the connection valves. A schematic description is depicted in Fig. 8.6.
The dynamic of the main tank needs to be extended with the flows induced by the new tank. Those flows depend on the height differences between the two levels, related to the static height of the lower and upper valves h_{32l} and h_{32u}:

$$\Delta_{2,3,l,k} = \max(h_{2,k}, h_{32l}) - \max(h_{3,k}, h_{32l}) \tag{8.8}$$

$$\Delta_{2,3,u,k} = \max(h_{2,k}, h_{32u}) - \max(h_{3,k}, h_{32u}). \tag{8.9}$$

The resulting model is again discretized using the Euler method with sampling time Δt

$$\begin{aligned} h_{2,k} = h_{2,k-1} &- \frac{1}{A_2} \underbrace{a_2\sqrt{2gh_{2,k-1}}\Delta t}_{\text{outflow}} - \frac{1}{A_2} \underbrace{\mathrm{sign}(\Delta_{2,3,l,k-1}) a_{32l} \sqrt{2g|\Delta_{2,3,l,k-1}|}\Delta t}_{\text{lower cross flow tank 3}} \\ &- \frac{1}{A_2} \underbrace{\mathrm{sign}(\Delta_{2,3,u,k-1}) a_{32u} \sqrt{2g|\Delta_{2,3,u,k-1}|}\Delta t}_{\text{upper cross flow tank 3}} + \frac{1}{A_2} \underbrace{\gamma_2 v_{2,k-1}\Delta t}_{\text{inflow pump 2}} + e_{2,k} \end{aligned} \tag{8.10}$$

with all values according to Appendix G, Tab. G.1.

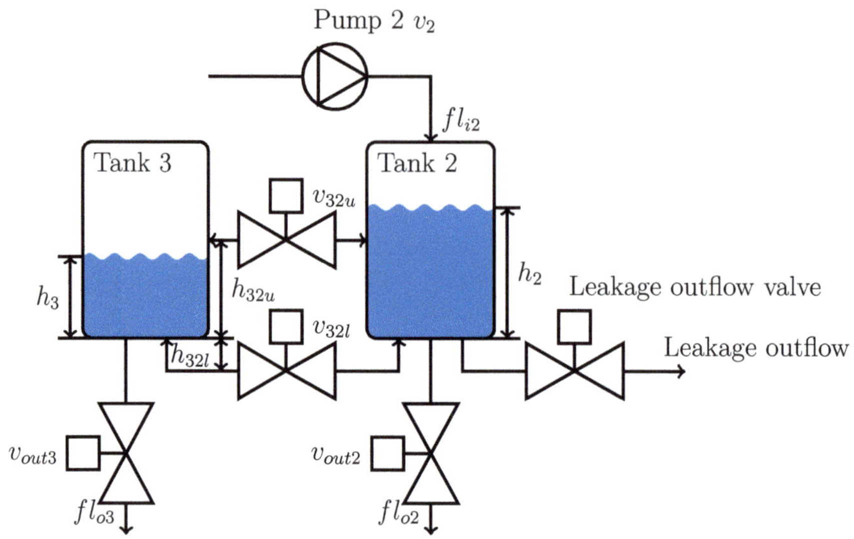

Figure 8.6: Schematic sketch of the two-tank experiment

The cross sections of both tanks are the same, i.e. $A_2 = A_3$, leading to symmetric height changes induced by the cross flow. The term $e_{2,k}$ is the additive disturbance including sensor and discretization fault. The system description can again be transformed to the pseudo linear regressor form (8.3) with

$$\varphi_k = \left[h_{2,k-1},\ |\Delta_{2,3,l,k-1}|,\ |\Delta_{2,3,u,k-1}|\right] \tag{8.11}$$

$$\Theta_k = \left[\theta_k^{(1)},\ \theta_k^{(2)},\ \theta_k^{(3)}\right]^T \tag{8.12}$$

$$y_k = h_{2,k} - \frac{\gamma_2}{A_2} v_{2,k-1} \Delta t. \tag{8.13}$$

The elements of the parameter vector Θ_k are given by

$$\theta_k^{(1)} = 1 - \frac{a_2}{A_2}\sqrt{\frac{2g}{h_{2,k-1}}}\Delta t + e'_{2,k} \tag{8.14}$$

$$\theta_k^{(2)} = \text{sign}(\Delta_{2,3,l,k-1})\frac{a_{32l}}{A_2}\sqrt{\frac{2g}{|\Delta_{2,3,l,k-1}|}}\Delta t + e''_{2,k} \tag{8.15}$$

$$\theta_k^{(3)} = \text{sign}(\Delta_{2,3,u,k-1})\frac{a_{32u}}{A_2}\sqrt{\frac{2g}{|\Delta_{2,3,u,k-1}|}}\Delta t + e'''_{2,k} \tag{8.16}$$

with unknown composition of the fault $e_{2,k} = {e'_{2,k}}/{h_{2,k-1}} + {e''_{2,k}}/{|\Delta_{2,3,l,k-1}|} + {e'''_{2,k}}/{|\Delta_{2,3,u,k-1}|}$. The parameters $\theta_k^{(2)}$ and $\theta_k^{(3)}$ given in (8.15) and (8.16) show singularities in case $|\Delta_{2,3,l,k-1}|$ or $|\Delta_{2,3,u,k-1}|$ approach zero.

Therefore the enclosing intervals $\boldsymbol{\theta}^{(2)}$ and $\boldsymbol{\theta}^{(3)}$ become very large when the operation range $[h_{2,min},\ h_{2,max}]$ is including or close to the height of the valve h_{32u}. It is thus necessary to chose the operation range with a sufficient distance to h_{32u} to achieve meaningful parameter intervals.

8.2.1 Measurement With Mapped State Signal

The first approach for the verification of the resulting hybrid system was introduced in Section 6.1 and used a so called "mapped state signal" according to Definition 6.10.
This means that the exact switching times and the respective active subsystems are known correctly. Therefore the hybrid verification task is reduced to sequential verification of the subsystems present in the measurement data.
The general system behavior given in equation (8.10) is therefore transferred to a more specific setting. Tank 3 is assumed to be empty ($h_{3,1} = 0\text{cm}$) with its nominal outflow valve open and the lower connection valve v_{32l} closed. The upper connection valve v_{32u} is open, as well as the nominal outflow valve of tank 2.
The resulting hybrid scenario consists of two states: State 1 is active if $h_{2,k} \leq h_{32u}$, i.e. the upper connection valve does not influence the system dynamics. State 2 is active for $h_{2,k} > h_{32u}$, i.e. an additional outflow is given through the upper connection valve.

The hybrid scenario is described as follows:
In state 1, starting at $h_{2,1} = 5\text{cm}$, pump 2 is used to fill tank 2. The nominal behavior of tank 2 is identical with the single-tank behavior described in Section 8.1.
State 2 is reached, when h_2 rises above the height of the upper connection valve, i.e. $h_{2,k} > h_{32u} = 30\text{cm}$. Thus the system dynamic changes due to the additional cross flow from tank 2 to tank 3 through the connection valve.

The resulting dynamic of state 2 is given by

$$h_{2,k} = h_{2,k-1} - \frac{1}{A_2}\left(\underbrace{u_2\sqrt{2gh_{2,k-1}}\Delta t}_{\text{nominal outflow}} + \underbrace{a_{32u}\sqrt{2g|h_{2,k-1} - h_{32u}|}\Delta t}_{\text{upper cross flow to tank 3}} - \underbrace{\gamma_2 v_{2,k-1}\Delta t}_{\text{inflow by pump 2}} + e_{2,k}\right). \tag{8.17}$$

In state 2 the level of tank 2 is always higher than the upper connection valve i.e. $h_{2,k-1} \geq h_{32u}$. Therefore the absolute value operator $|\cdot|$ on $(h_{2,k-1} - h_{32u})$ can be dropped and the singularity present in (8.16) is avoided.

The upper cross flow to tank 3, given in (8.17) can thus be reformulated:

$$a_{32u}\sqrt{2g(h_{2,k-1} - h_{32u})}\Delta t \tag{8.18}$$

$$= a_{32u}(h_{2,k-1} - h_{32u})\sqrt{\frac{2g}{(h_{2,k-1} - h_{32u})}}\Delta t \tag{8.19}$$

$$= a_{32u}h_{2,k-1}\sqrt{\frac{2g}{(h_{2,k-1} - h_{32u})}}\Delta t - a_{32u}h_{32u}\sqrt{\frac{2g}{(h_{2,k-1} - h_{32u})}}\Delta t \tag{8.20}$$

$$= h_{2,k-1}\left(a_{32u}\sqrt{\frac{2g}{(h_{2,k-1} - h_{32u})}}\Delta t - a_{32u}h_{32u}\sqrt{\frac{2g}{h_{2,k-1}^2(h_{2,k-1} - h_{32u})}}\Delta t\right) \tag{8.21}$$

Finally, the model for state 2 is given in in pseudo linear regressor form (8.3) with

$$\varphi_k = h_{2,k-1} \tag{8.22}$$

$$\theta_k = 1 - \frac{a_2}{A_2}\sqrt{\frac{2g}{h_{2,k-1}}}\Delta t - \frac{a_{32u}}{A_2}\sqrt{\frac{2g}{(h_{2,k-1} - h_{32u})}}\Delta t + \frac{a_{32u}}{A_2}h_{32u}\sqrt{\frac{2g}{h_{2,k-1}^2(h_{2,k-1} - h_{32u})}}\Delta t + \frac{e_{2,k}}{h_{2,k-1}} \tag{8.23}$$

$$y_k = h_{2,k} - \frac{\gamma_2}{A_2}v_{2,k-1}\Delta t. \tag{8.24}$$

The nominal parameter θ_k can again be determined depending on the level $h_{2,k-1}$ for different outflow configurations. The resulting parameter ranges are depicted in Fig. 8.7. and are used to determine the nominal values of the system parameters.

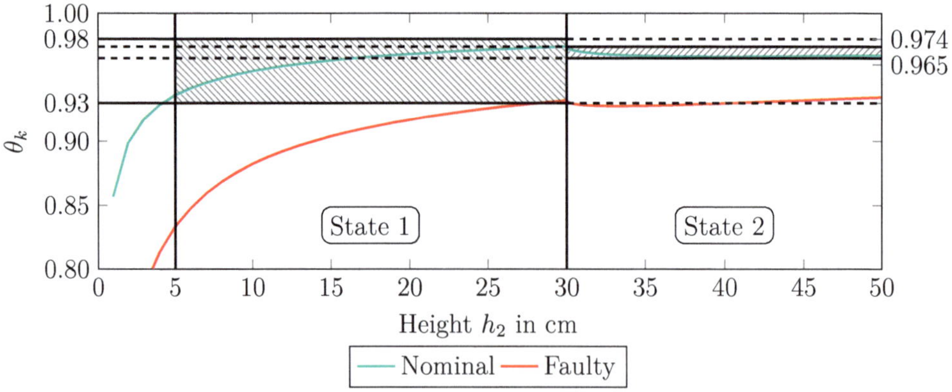

Figure 8.7: Values of θ_k depending on h_2 for different outflow configurations at state 1 (left) and state 2 (right)

State 1 is assigned with the same values as in Section 8.1 i.e.

$$\boldsymbol{\theta}(1)^* = [0.93,\ 0.98]\,. \tag{8.25}$$

In state 2, i.e. starting from $h_{2,k-1} = 30\text{cm}$, the time variant parameter for nominal outflow only can be enclosed by the interval

$$\boldsymbol{\theta}(2)^* = [0.965,\ 0.974]\,. \tag{8.26}$$

The nominal parameters are now used to verify the hybrid setting. Again the collected measurement data of the height h_2 is enclosed by interval values

$$\boldsymbol{h}_{2,k} = \left[h_{2,k}(1-\delta^r_{h_2}),\ h_{2,k}(1+\delta^r_{h_2})\right] \tag{8.27}$$

with $\delta^r_{h_2} = 0.15$.

The upper part of Fig. 8.8 shows a hybrid test run, starting with tank 3 being empty and the level of tank 2 being at $h_{2,1} = 5\text{cm}$. The level of tank 2 is increased using pump 2 and thus crosses the valve height h_{32u} at $k = 64$ seconds. The tank dynamics are changed by the crossing as the upper connection valve now acts as an additional outflow of tank 2.
The verification results are depicted in the lower part of Fig. 8.8. State 1 can be verified as long as the level in tank 2 is below the connection valve, i.e. $h_{2,k} \leq h_{32u}$. Afterwards, state 2 is verified until the end of the measurement sequence.

Note that in this case it is not necessary to perform cross validation, i.e. applying the parameters of state 2 to measurement from state 1 and vice versa. This is due to the fact that the real switching time and the respective active state are given by the mapped state signal.

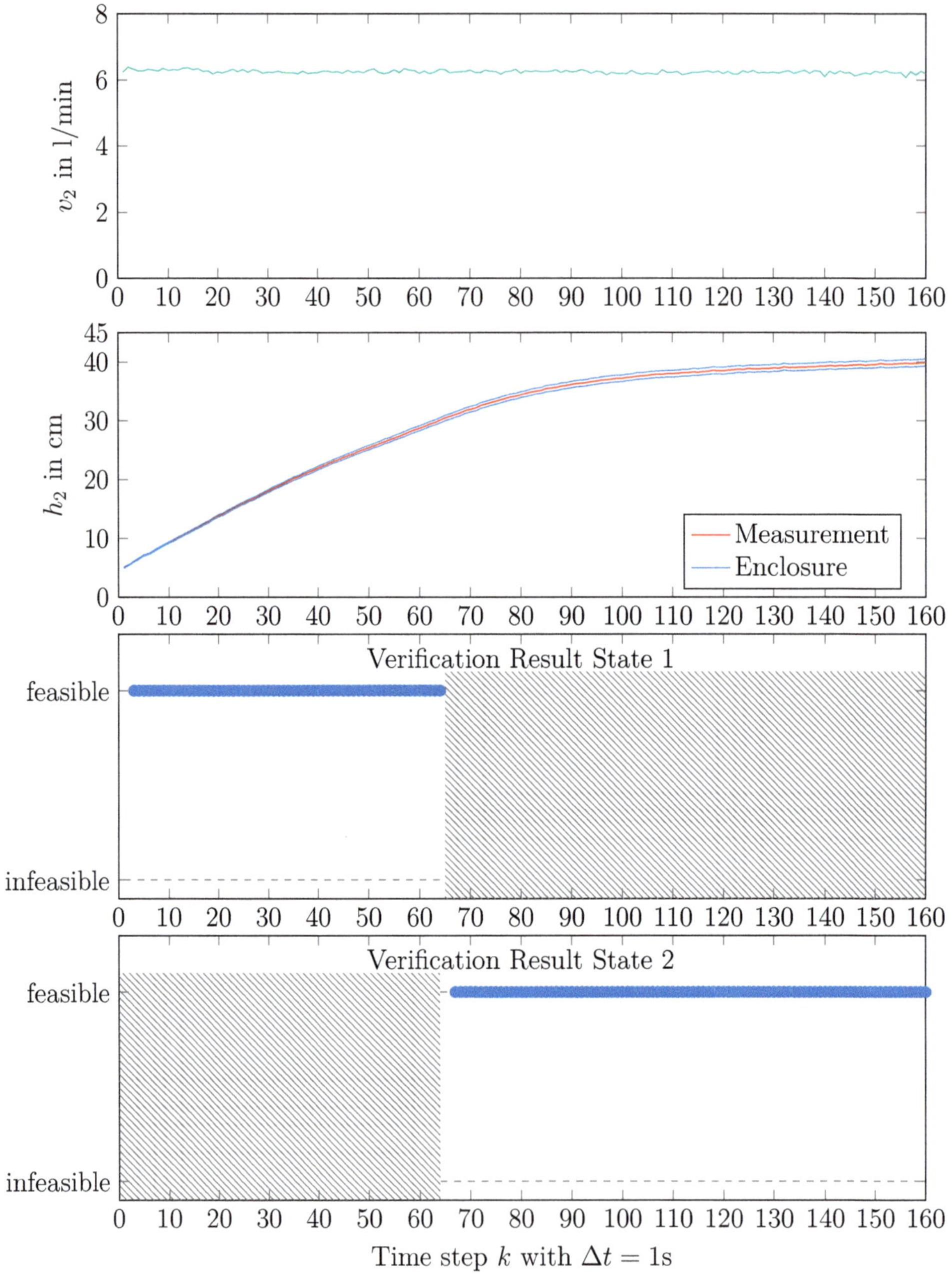

Figure 8.8: Verification result in the hybrid case for state 1 and state 2 with a mapped state signal

8.2.2 Measurement Without Mapped State Signal

The introduced setting is now generalized by omitting the mapped state signal, still assuming known switching times. This means that it is still known there is a state change at $k = 64$ in the scenario, but now it is unknown whether the system switches from state 1 to state 2 or vice versa. Therefore all segments have to be analyzed twice, using the nominal parameters of both states. This cross verification is used to determine the active state of the respective subsystem.
To ensure successful cross validation, Prager-Oettli-Distinguishability as given in Assumption 6.2 has to be fulfilled. Therefore the nominal parameters of state 1 and state 2 are investigated. It is obvious that the nominal parameters of state 2 are a subset of the nominal parameters of state 1, as

$$\boldsymbol{\theta}^*(1) = [0.93,\ 0.98] \supset [0.965,\ 0.974] = \boldsymbol{\theta}^*(2) \tag{8.28}$$

which leads to the existence of a common parameter

$$\boldsymbol{\theta}^*_{com} = \boldsymbol{\theta}^*(1) \cap \boldsymbol{\theta}^*(2) = [0.965,\ 0.974]\,. \tag{8.29}$$

Therefore all parameters that are within the set of common parameters $\boldsymbol{\theta}^*_{com}$ are consistent with both states by definition.
It is hence not possible to distinguish the states as Prager-Oettli-Distinguishability is not fulfilled in the current setting. Since this property is the main preliminary of hybrid verification without mapped state signal it is formally impossible to demonstrate the viability of the method using this setting.

The performance of the algorithm in case of fulfilled Prager-Oettli-Distinguishability was demonstrated in Example 6.2. Furthermore it was shown that it is possible to segment and verify the measurement data even without information about the switches if Prager-Oettli-Segmentability holds. The respective setting and the results are given in Example 6.3.

8.3 Simulation: Diagnosis By Kaucher Based Guaranteed Verification (Four-Tank)

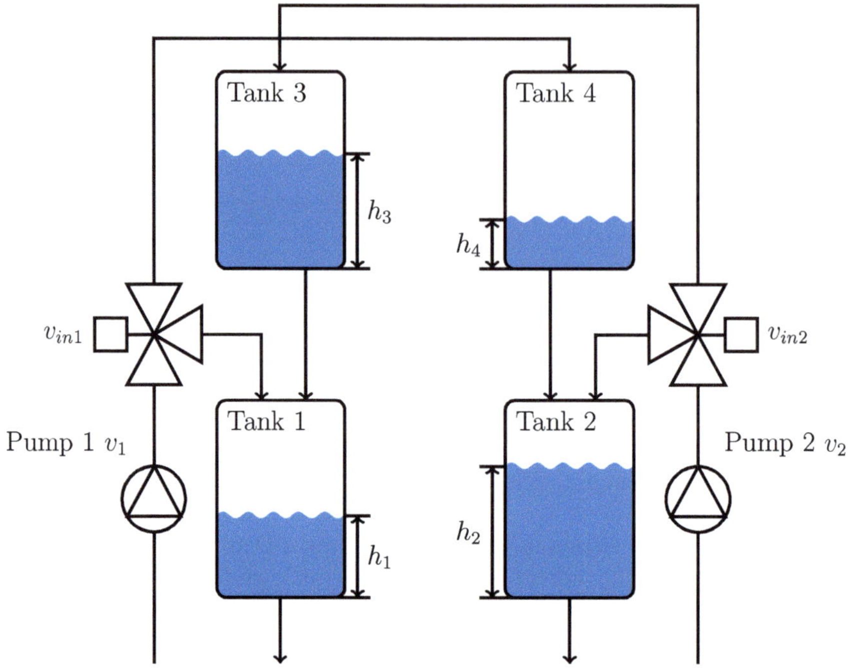

Figure 8.9: Schematic view of the used four-tank system

The regarded setting is now changed to a slightly different four-tank setup. The four-tank process is an established benchmark in literature and was proposed by [Joh00]. Here it is used to apply the diagnosis method introduced in Chapter 7. This application was published and presented in [Sch18b].

The four-tank setting is depicted in Fig. 8.9. For symmetry reasons the setting can be reduced to tank 1 and tank 3.

The dynamic of tank 1 is chosen to be the objective of the verification. The heights h_1 and h_3 of both tanks are measured, as well as the on/off signal v_1 of pump 1. The flow of pump 1 is split by the input valve v_{in1} leading to $v_{in1} = 0.7$ of the flow going to tank 1 and $(1 - v_{in1})$ of the flow going to tank 4. The flows from pump 1 as well as from tank 3 are considered as inputs. The respective equations are similar to (8.2), now taking into account an additional inflow depending on h_3.

All simplifications and the resulting model equations are based on the considerations of [Ble11].

Discretization using Euler Method and sampling time Δt leads to

$$h_{1,k} = h_{1,k-1} - \frac{a_1}{A_1} \underbrace{a_1 \sqrt{2gh_{1,k-1}} \Delta t}_{\text{outflow}} + \frac{1}{A_3} \underbrace{a_3 \sqrt{2gh_{3,k-1}} \Delta t}_{\text{inflow from tank 3}} + \frac{1}{A_1} \underbrace{v_{in1} \gamma_1 v_{1,k-1} \Delta t}_{\text{inflow by pump 1}} + e_{1,k} \tag{8.30}$$

with parameters according to Appendix G, Tab. G.2. The additive disturbance $e_{1,k}$ includes sensor and discretization faults.
The pseudo linear regressor form (8.3) is now given by

$$y_k = h_{1,k} - \frac{v_{in1}\gamma_1}{A_1} v_{1,k-1} \Delta t \tag{8.31}$$

$$\varphi_k = [h_{1,k-1} \; h_{3,k-1}] \tag{8.32}$$

$$\Theta_k = \left[\theta_k^{(1)} \; \theta_k^{(2)}\right]^T \tag{8.33}$$

with the time variant parameters

$$\theta_k^{(1)} = 1 - \frac{a_1}{A_1} \sqrt{\frac{2g}{h_{1,k-1}}} \Delta t \tag{8.34}$$

$$\theta_k^{(2)} = \frac{a_3}{A_3} \sqrt{\frac{2g}{h_{3,k-1}}} \Delta t. \tag{8.35}$$

It is assumed that the operation range of the tank system is $h_{1,k} \in [2, \; 10.5]$ cm and $h_{3,k} \in [1, \; 15]$ cm which leads to $\boldsymbol{\theta}^{(1)*} = [0.921, \; 0.965]$ and $\boldsymbol{\theta}^{(2)*} = [0.029, \; 0.112]$.
The resulting midpoint radius expressions of the parameters are $\theta_c^{(1)*} = 0.943$, $\theta_\Delta^{(1)*} = 0.022$ and $\theta_c^{(2)*} = 0.0705$, $\theta_\Delta^{(2)*} = 0.0415$.
The optimization based diagnosis approach using a zonotopic approximation is chosen in this example. Therefore the measurement data and the nominal parameter set are used to set up the constraints of the optimization problem.
The nominal feasible parameter box $\boldsymbol{\Theta}^*$ is used to build the initial zonotope with:

$$P^0 = \begin{bmatrix} \theta_c^{(1)*} & \theta_c^{(2)*} \end{bmatrix}^T \tag{8.36}$$

$$H^0 = \begin{bmatrix} \theta_\Delta^{(1)*} & 0 \\ 0 & \theta_\Delta^{(2)*} \end{bmatrix}. \tag{8.37}$$

Afterwards the outer enclosure of the intersection between initial zonotope and measurement data is calculated using (7.10)-(7.12). The resulting zonotope is used as starting point P_0^0, H_0^0 of the optimization problem. The solution of the optimization problem is thus a zonotopic approximation of the united solution set given by $\mathcal{Z}$. All parameter vectors included in the optimal solution set $\mathcal{Z}$ are solutions of the ILES (3.46). If the intersection of the specification and measurement is nonempty, the algorithm calculates a feasible set in this area. The results and limitations of the approach are demonstrated in the following, using several different settings.

8.3.1 Fault Free Setting

First a fault free scenario is given as depicted in Fig. 8.10. The scenario includes parts with pump on and off and thus shows a variety of different water level dynamics both in tank 1 and tank 3. The measurement data of h_1 and h_3 are enclosed using intervals with radius $\delta^a_{h_1} = \delta^a_{h_3} = 0.05\text{cm}$ leading to the interval values

$$\boldsymbol{h}_{1,k} = \left[h_{1,k} - \delta^a_{h_1},\ h_{1,k} + \delta^a_{h_1}\right] \tag{8.38}$$

$$\boldsymbol{h}_{3,k} = \left[h_{3,k} - \delta^a_{h_3},\ h_{3,k} + \delta^a_{h_3}\right]. \tag{8.39}$$

The results of the optimization based zonotopic method are given in subplot 4 of Fig. 8.10. The algorithm calculated a feasible set of parameters for the time segments $[1,\ k_{end}]$ with $k_{end} \in [[1,\ 1297],\ [1572,\ 2000]]$. However, time segments starting in the beginning and ending in $k_{end} \in [1298,\ 1571]$ are not verified. Therefore there is temporarily no consistency according to Prop. 7.4 given in this segment. This is due to the CMP effect as introduced in Section 7.2.1.
A detailed view on the relevant time instants is given in Fig. 8.11 which displays the change of the result from consistent to inconsistent and back.

The constraints given by the measurement data are depicted by the blue lines in Fig. 8.11, the nominal set by the shaded area and the zonotopic approximation of the united solution set is shown as the green zonotope.
At $k = 1297$, the measurement data is proven to be basic consistent with the specification as there is a nonempty intersection between nominal set and the approximation of the united solution set (orange part of the zonotope in Fig. 8.11, subfigure 1).
At $k = 1298$, the verification result is inconsistent for the first time. However, there is still a feasible region within the nominal parameter set, shown by the green crosses that depict vertexes fulfilling Prop. 4.1 (Fig. 8.11, subfigure 2). Those points were used by the algorithm while calculating a suitable factor α. However there is no intersection between the final green zonotope and the shaded nominal region. This behavior reflects exactly the definition of the CMP effect introduced in Section 7.2.1. The CMP effect is observable until $k = 1571$, (Fig. 8.11, subfigure 3).
Starting from $k = 1572$, additional constraints given by new measurement data are taken into account. Therefore, the center and shape of the outer enclosure is changed. This results in a zonotopic approximation of the united solution set that provides a nonempty intersection with the nominal set again (Fig. 8.11, subfigure 4), leading to a successful verification.
As those later results are calculated based on all measurement data, including the possibly inconsistent times $k \in [1298,\ 1571]$, the results showing the CMP effect are corrected and the verification result for $k > 1571$ can be generalized for all $k \in [1, 2000]$.

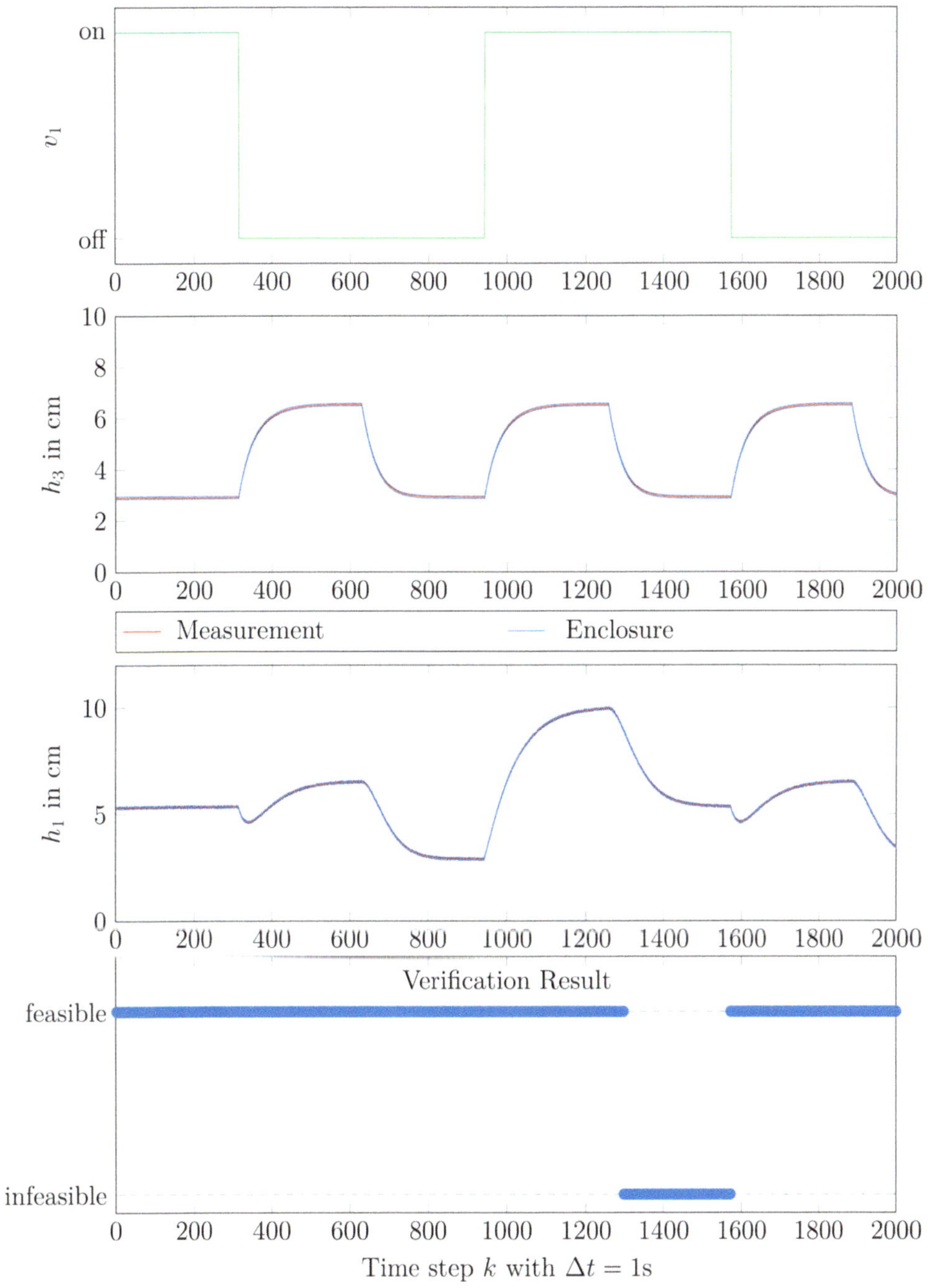

Figure 8.10: Verification result for the fault free setting

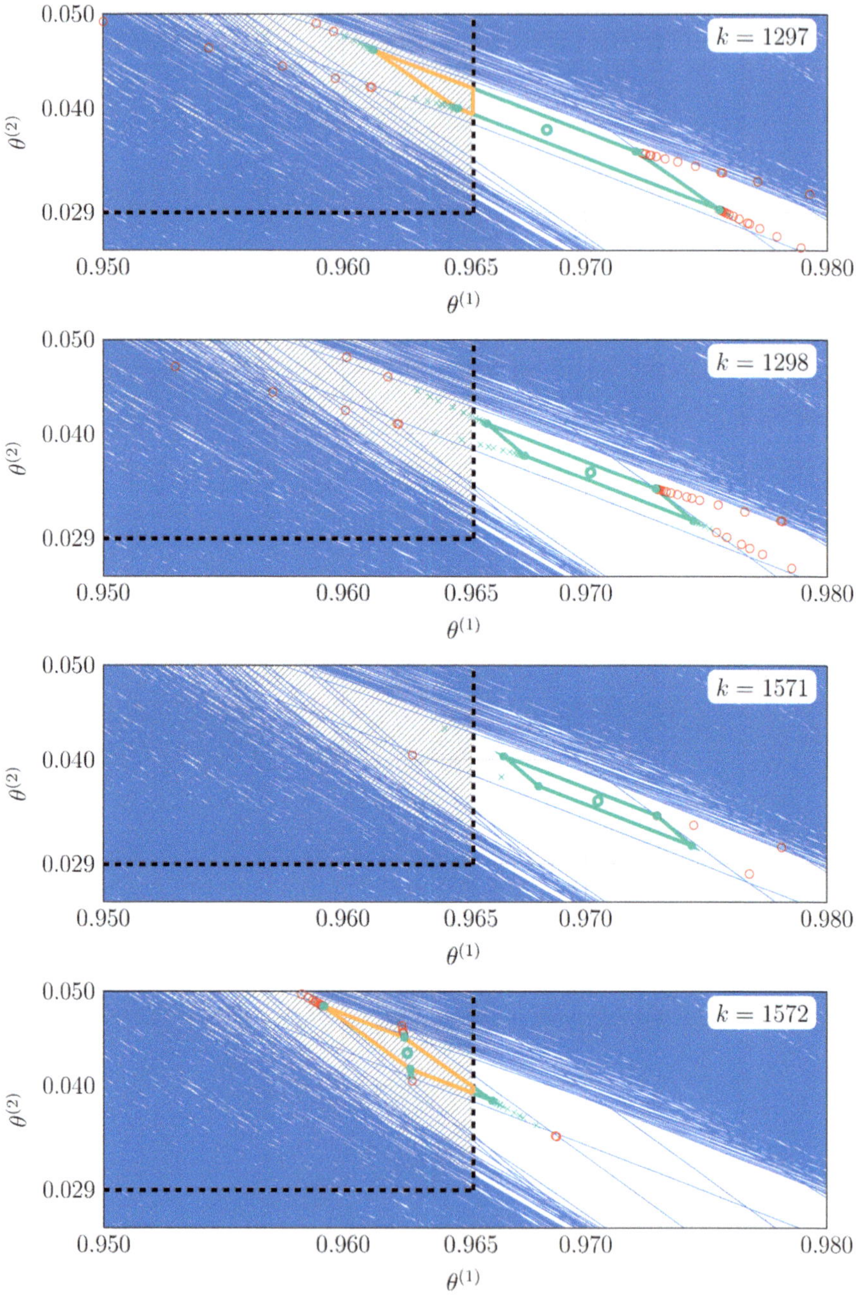

Figure 8.11: Zoom on the time instants showing the CMP effect

The method is applied iteratively to calculate an individual verification result for each time step k. The calculations are done offline and the necessary calculation time for the complete data set is $T_{calc} = 703.8\text{s} < 2000\text{s} = T$.
Regarding the calculation time of each measurement time step shows that $T_{calc,k} < 1\text{s}$ which means that the method might in general be real time capable. However, the algorithm is based on an optimization procedure with non-deterministic runtime. Therefore special measures need to be taken to ensure deterministic runtime of each step. For an initial estimate the average runtime of the optimization algorithm is used by regarding the total runtime of the method throughout this chapter.

8.3.2 Additive Faults

There are several phenomena that can lead to an additive fault of an sensor. A possible sensor fault is called "freeze", when the sensor will return a fixed constant value. Another common sensor fault is "offset", which means that the sensor will add a constant bias to the true measurement value. A third additive sensor property is the specific sensor noise. Noise is not regarded as an effect to be detected here. However it is crucial to know the sensor noise precisely to choose the bound of the interval enclosure correctly.

Given the correct faultless but noisy sensor data s_k, a freeze fault of value f_f, occurring at time k_{err}, can be expressed as follows:

$$s_{f,k} = \begin{cases} s_k & \forall k \in [1, k_{err} - 1] \\ f_f & \forall k \in [k_{err}, T] \,. \end{cases} \tag{8.40}$$

The measurement values of h_1 and h_3 are enclosed using $\delta^a_{h_1} = \delta^a_{h_3} = 0.05\text{cm}$. The case of a freeze fault of $f_f = 6.0\text{cm}$ on measurement h_1 at $k_{err} = 650$ is depicted in Fig. 8.12. It can be seen that the detection time is equal to the fault time $k_{det} = 650 = k_{err}$ which means that the freeze fault is detected instantaneously.
The results for several different freeze fault amplitudes on h_1 and the respective fault detection times k_{det} are given in Tab. 8.1. All detected faults were checked in detail to identify settings showing the CMP effect.
In case of $f_f = 6.2\text{cm}$ a CMP condition occurred. This is due to the fact that the used freeze fault intensity is enclosed in the $\delta^a_{h_1} = 0.05\text{cm}$ interval around the real value of $h_{1,650} = 6.22\text{cm}$. Thus at $k = 651$ the freeze fault case is really close to the correct value of $h_{1,650+1}$, meaning that there is a feasible parameter mapping $h_{1,k} = 6.2$ to $h_{1,k+1} = 6.2$. Subsequent points do not provide additional information, as the sensor value is fixed by the freeze fault. The only additional information is provided by the pump signal v_1. The varying pump signal leads to a movement of the center of the outer enclosing zonotope. At $k_{det} = 684$ the center of the zonotope is moved to a position that generates a CMP effect that is erroneously interpreted as an inconsistency.

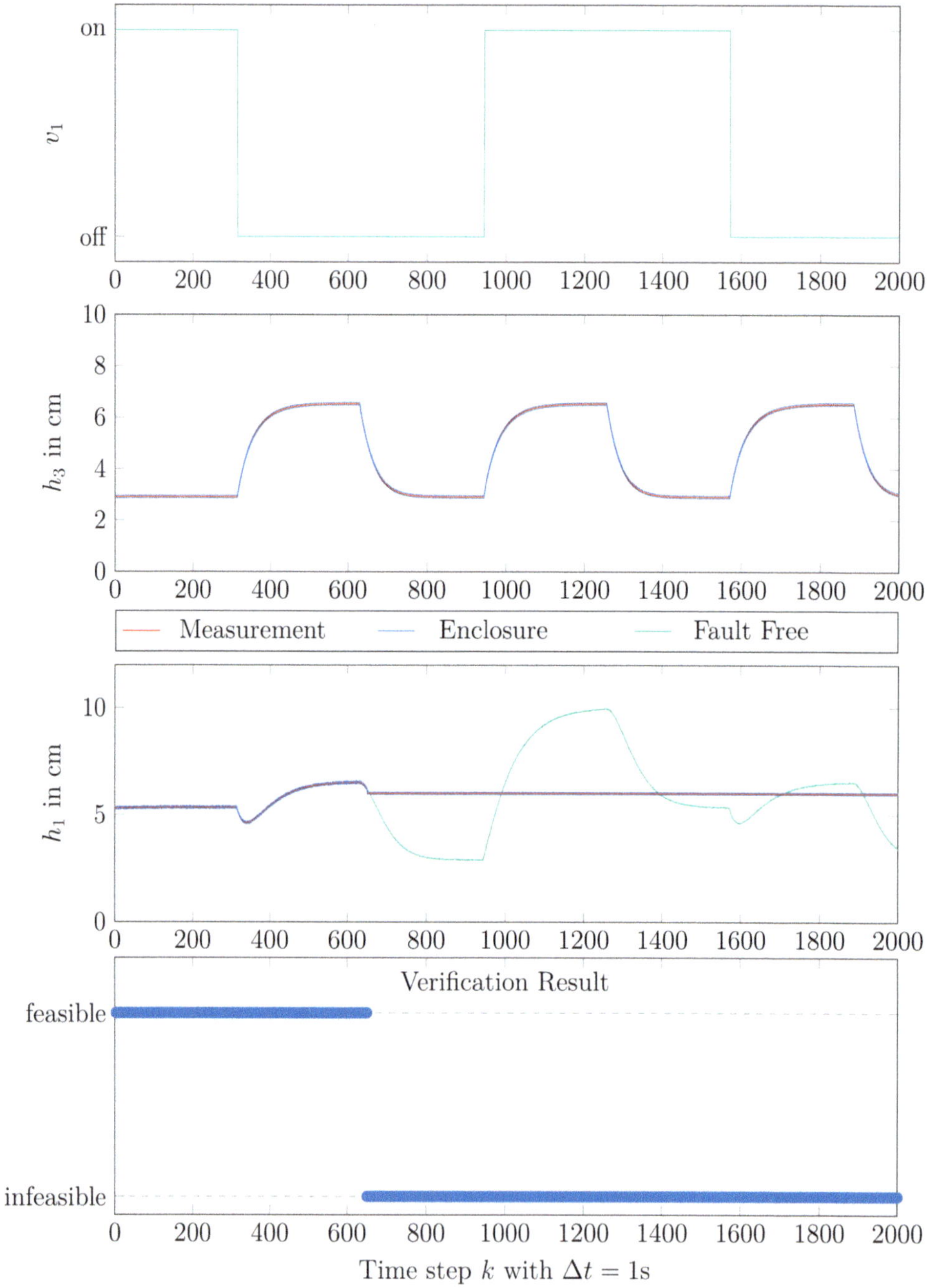

Figure 8.12: Verification results for freeze fault of $f_f = 6.0$cm at $k_{err} = 650$

All calculation times for freeze faults given in Table 8.1 are less than the time of the measurement signal $T_{calc} < 2000\text{s} = T$.

Table 8.1: Different fault amplitudes and resulting detection times for freeze fault

Fault f_f in cm	Time step of Fault k_{err}	Time step of Detection k_{det}	Quality	Calculation Time T_{calc} in s
7.0	650	650	no CMP	1181.0
6.5	650	650	no CMP	1200.9
6.2	650	684	CMP occurred	743.4
6.0	650	650	no CMP	1461.7

The second regarded malfunction is an offset fault. In this case the sensor value is not fixed, but a specific value is added to each measurement:

$$s_{o,k} = \begin{cases} s_k & \forall k \in [1, k_{err} - 1] \\ s_k + f_o & \forall k \in [k_{err}, T] \, . \end{cases} \tag{8.41}$$

Again the measurement values of h_1 and h_3 are enclosed using $\delta^a_{h_1} = \delta^a_{h_3} = 0.05\text{cm}$. The results for an offset fault of $f_o = 0.7\text{cm}$ on sensor h_1 are depicted in Fig. 8.13. Again the verification result changes to infeasible right at the moment the fault gets effective i.e., at $k_{det} = 650 = k_{err}$.

Further results for different fault amplitudes are given in Table 8.2. Instantaneous detection is possible up to a fault amplitude of $f_o = 0.15\text{cm}$. Note that the measurement noise is enclosed using $\delta^a_{h_1} = \delta^a_{h_3} = 0.05\text{cm}$ which leads to an interval width of $2\delta^a_{h_1} = 0.1\text{cm}$. This is very close to the fault amplitude $f_o = 0.15\text{cm}$. When using $f_o = 0.1\text{cm}$ - which is exactly the interval width - the CMP effect occurs. The necessary calculation time is less than the signal duration for all regarded offset fault intensities.

Table 8.2: Different fault amplitudes and resulting detection times for offset fault

Fault f_o in cm	Time step of Fault k_{err}	Time step of Detection k_{det}	Quality	Calculation Time T_{calc} in s
0.70	650	650	no CMP	1367.9
0.30	650	650	no CMP	1475.0
0.20	650	650	no CMP	1361.5
0.15	650	650	no CMP	1210.0
0.10	650	650	CMP occurred	1036.8

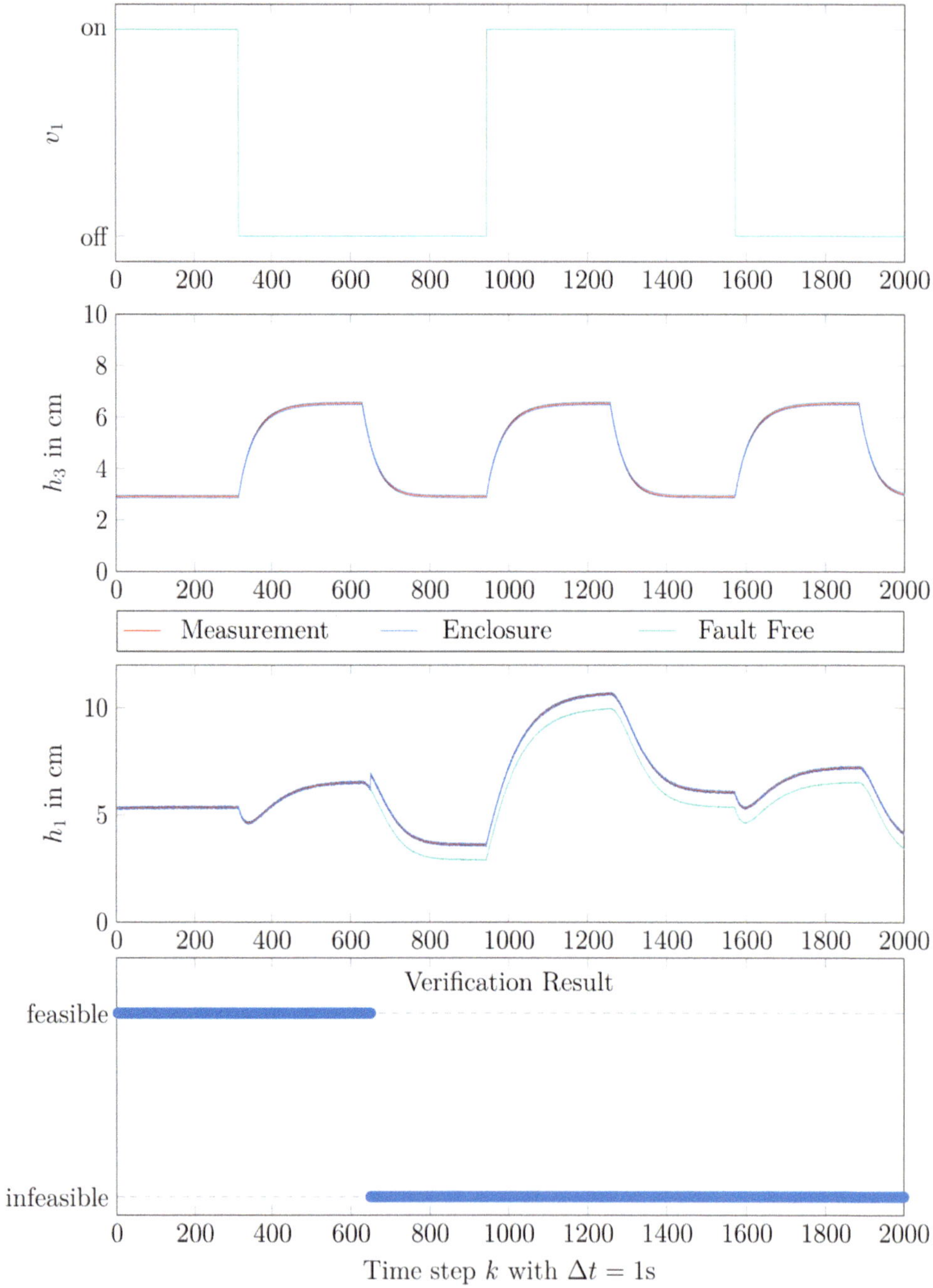

Figure 8.13: Verification results for offset fault of $f_o = 0.7\text{cm}$ at $k_{err} = 650$

8.3.3 Multiplicative Faults

Multiplicative faults can be related to faults in system components i.e. a congested or leaking pipe or decreasing pump performance. Such a multiplicative fault f_θ directly influences the system parameter:

$$\theta_{err,k} = \begin{cases} \theta_k & \forall k \in [1, k_{err} - 1] \\ \theta_k + f_\theta & \forall k \in [k_{err}, T] \, . \end{cases} \tag{8.42}$$

A maximum absolute deviation of $\delta^a_{h_1} = \delta^a_{h_3} = 0.05\text{cm}$ is used to enclose the measurement values of h_1 and h_3. An exemplary setting for $f_{\theta^{(1)}} = 0.035$ at $k_{err} = 1200$ is depicted in Fig. 8.14, further results are given in Tab. 8.3.
It can be seen, that the faulty parameter influences the value of h_1. This change in system dynamic is recognized by the verification method at $k_{det} = 1200 = k_{err}$.
All detected inconsistencies were checked in detail. Reliable results are possible up to $f_{\theta_1} = 0.010$. This is a very small value with respect to the nominal parameter variability $\theta^{(1)}_\Delta = 0.025$ which shows the new method is very sensitive.
The condition $T_{calc} < 2000\text{s} = T$ holds for all entries in Tab. 8.3.

Table 8.3: Different fault amplitudes and resulting detection times for parameter fault

Fault $f_{\theta^{(1)}}$	Time step of Fault k_{err}	Time step of Detection k_{det}	Quality	Calculation Time T_{calc} in s
0.035	1200	1200	no CMP	1036.6
0.022	1200	1215	no CMP	1123.9
0.020	1200	1227	no CMP	1166.7
0.010	1200	1572	no CMP	764.7
0.005	1200	1638	CMP occurred	746.7

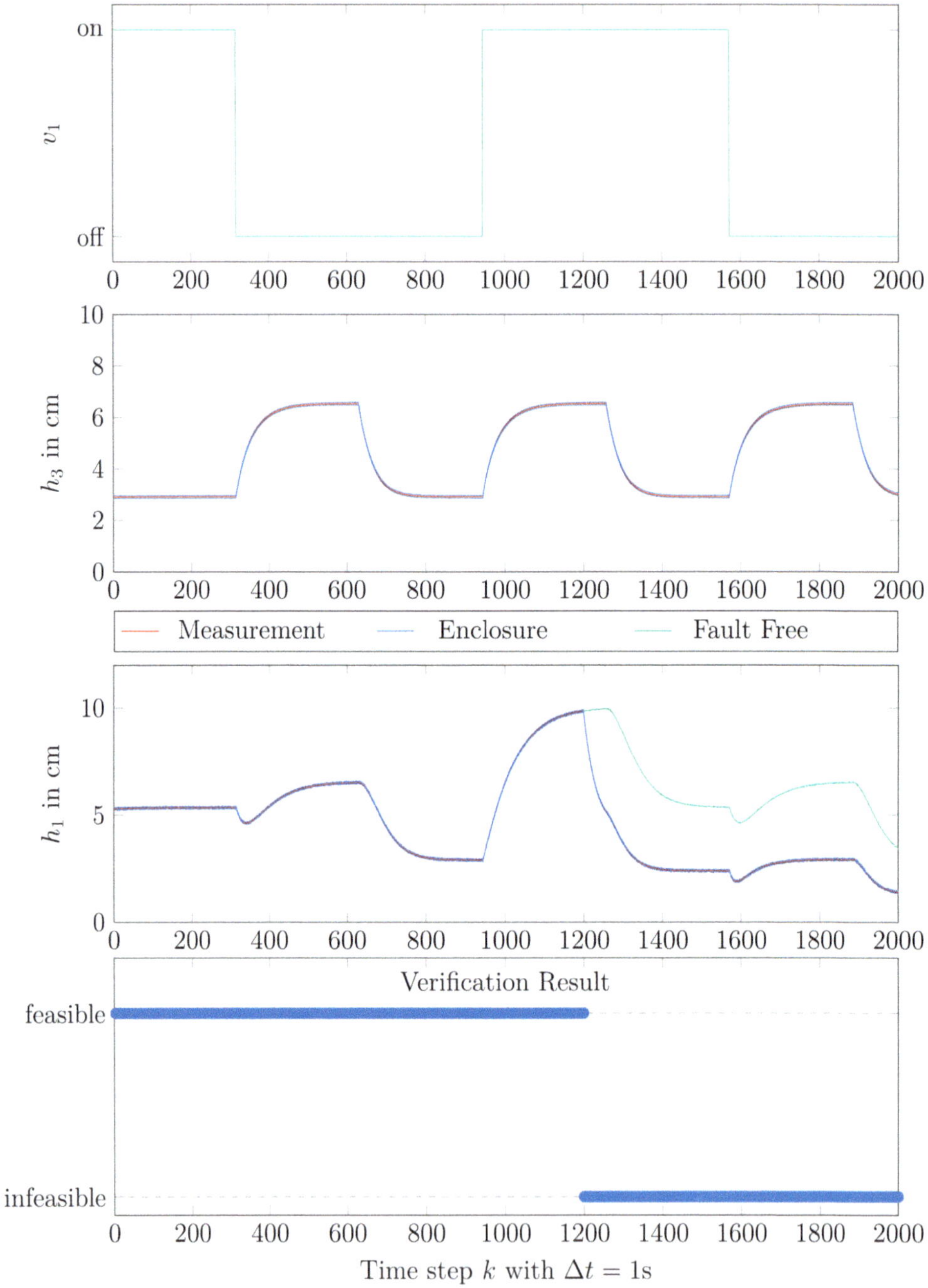

Figure 8.14: Verification results for multiplicative fault of $f_{\theta(1)} = 0.035$ at $k_{err} = 1200$

8.4 Application: Diagnosis By Kaucher Based Guaranteed Verification (Single-Tank)

The diagnosis method is now applied to real measurement data instead of simulation data as in the previous chapter. Therefore the IRS three-tank setting (introduced in Section 8.1) is used, again reduced to the single-tank setup. The respective geometric parameters can be taken from Appendix G, Tab. G.1.
The following scenario is regarded: The water level in tank 2 has an initial height of $h_{2,1} = 24.48\text{cm}$ and is rising due to the input flow from pump 2. Pump 2 is running at a high load with varying intensity.
First, the fault free setting is evaluated to show that the method is able to verify the nominal setting. Then the additive sensor faults "freeze" and "offset" are applied to the measurement data. Finally a scaling fault on the height measurement data is considered.

The results of different fault intensities as well as the detection and calculation times are given in several tables. Each result was evaluated carefully to determine CMP conditions. The results were initially published and presented in [Sch18c].

8.4.1 Fault Free Setting

The regarded operation range is defined to be $h_2 \in [24,\ 46]\ \text{cm}$ and used with (8.5) to obtain the nominal range $\boldsymbol{\theta}^* = [0.971,\ 0.979]$.
The result is calculated by using an interval width of $\delta^a_{h_2} = 0.4\text{cm}$ to enclose the measurement data of h_2. The pump measurement data is assumed to be noiseless and thus used as point real value.
The fault free behavior is depicted in Fig. 8.15. It is verified for the entire measurement time. The necessary calculation time is $T_{calc} = 34.4\text{s} < 120\text{s} = T$.

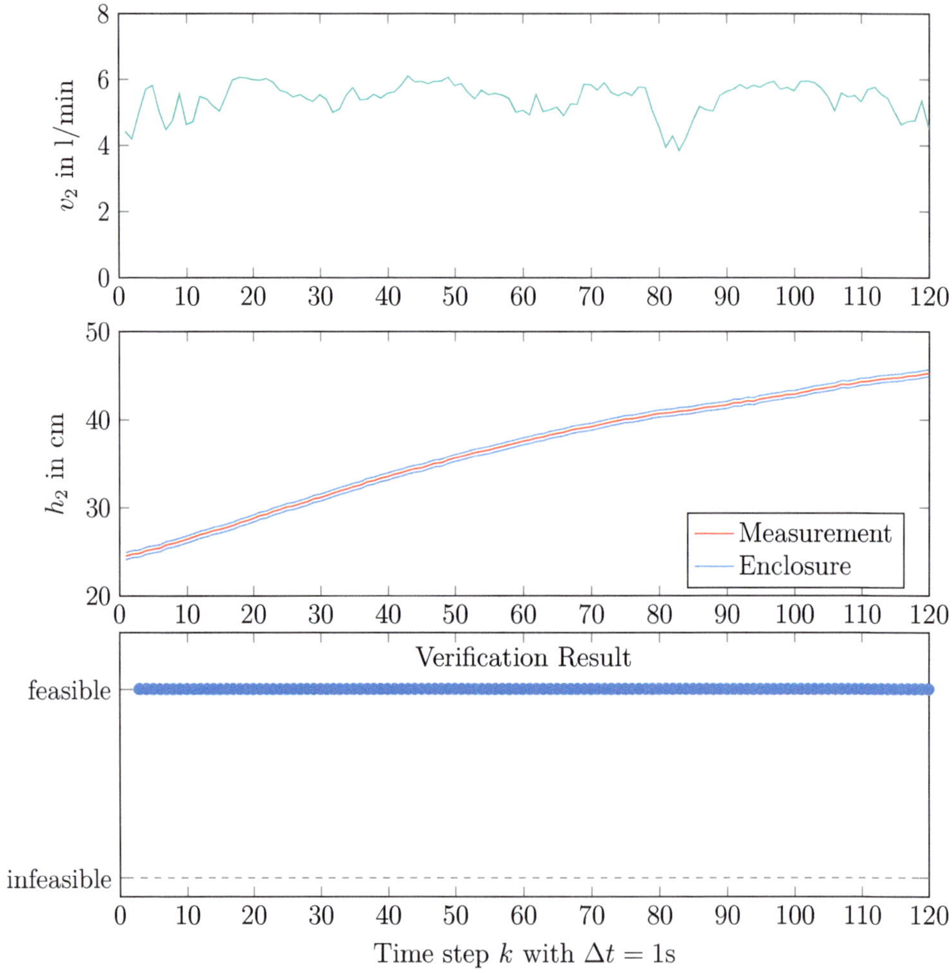

Figure 8.15: Fault free measurement data of the single-tank diagnosis scenario

8.4.2 Additive Faults

The first considered additive fault is given by a sensor freeze. The used mathematical model to distort the fault free measurement data of h_2 is given by (8.40). The measurement is enclosed using an absolute deviation of $\delta^a_{h_2} = 0.4\text{cm}$. The resulting system run for $f_f = 37.9\text{cm}$ on the measurement of h_2 at $k_{err} = 60$ is depicted in Fig. 8.16.
It can be seen that it is not possible to verify the measurement data as soon as the freeze fault is active. The failure is detected at $k_{det} = 60 = k_{err}$, i.e. at the very first time the measurement is distorted.
An evaluation of the performance of the method for several different freeze fault intensities f_f is listed in Tab. 8.4. It can be seen that it is possible to detect faults in a large range from $f_f = 42.0\text{cm}$ to $f_f = 37.9\text{cm}$. The lower value is very close to the correct value $h_{2,k_{err}} = 37.54\text{cm}$.
All faults are detected directly at their first appearance, i.e. at $k_{det} = 60 = k_{err}$. All results were checked carefully to ensure that there is no CMP effect present in the results.
All calculation times are less than the measurement time, i.e. $T_{calc} < 120\text{s} = T$.

Table 8.4: Different freeze fault amplitudes for $s_{k_{err}} = h_{2,60} = 37.54\text{cm}$

Fault f_f in cm	Time step of Fault k_{err}	Time step of Detection k_{det}	Quality	Calculation Time T_{calc} in s
42.0	60	60	no CMP	99.6
39.0	60	60	no CMP	49.6
38.5	60	60	no CMP	50.2
38.0	60	60	no CMP	80.8
37.9	60	60	no CMP	31.0
37.7	60	not detected	no CMP	24.9

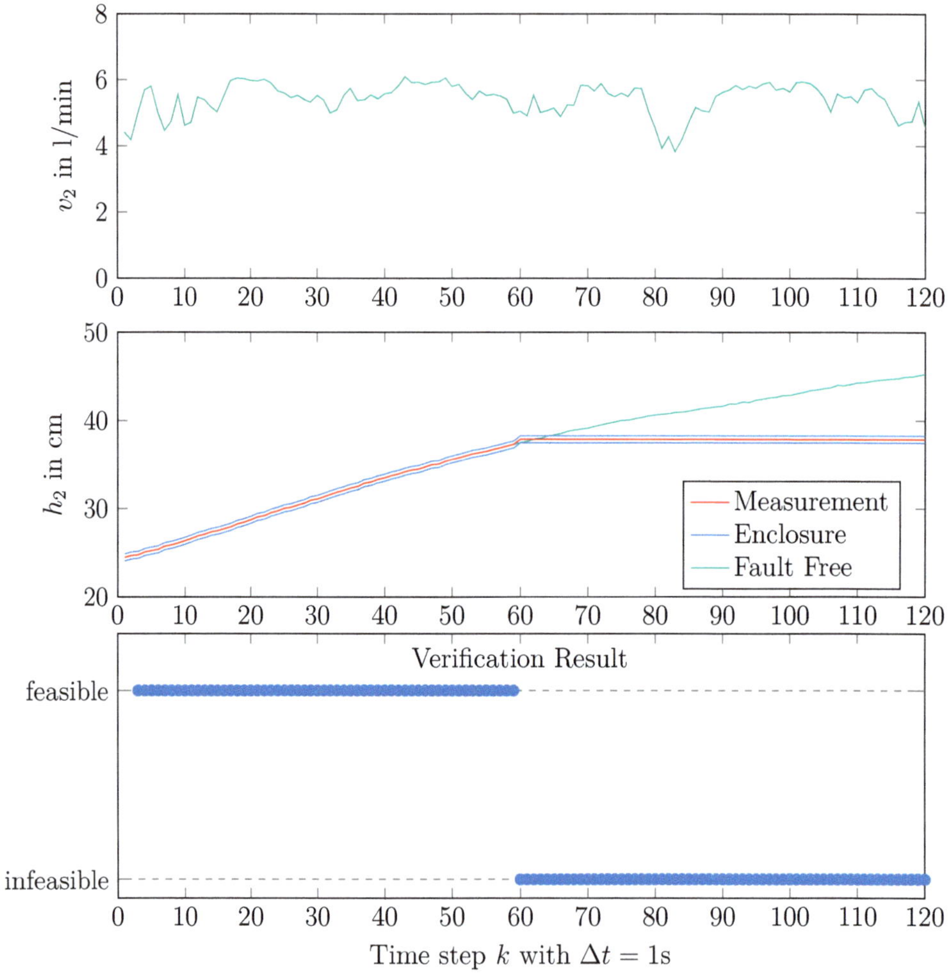

Figure 8.16: Measurement data with freeze fault $f_f = 37.9$cm

Second, an offset fault setting is applied. Therefore a constant offset f_o is added to the faultless measurement data of h_2, according to (8.41). Again, an absolute deviation of $\delta^a_{h_2} = 0.4\text{cm}$ is used to enclose the measurement. The performance of the method is shown exemplary in Fig. 8.17 and Fig. 8.18.
The large offset of $f_o = 5\text{cm}$ in Fig. 8.17 is rather obvious and could also be detected by an expert. On the other hand, the very small offset of $f_o = 0.35\text{cm}$ in Fig. 8.18 is very hard to distinguish from the fault free measurement depicted in green.
Nevertheless the zonotopic method is able to detect it at the moment of its first appearance. This is a very powerful property as the detected offset of $f_o = 0.35\text{cm}$ is smaller than the used interval radius of $\delta^a_{h_2} = 0.4\text{cm}$.
This performance is due to the dynamic between time instant $k_{err} - 1$ and time instant k_{err} that is created by the appearance of the offset fault. This dynamic is detected instantaneously as it is outside of the nominal parameter range.

Several results for different offset intensities are given in Tab. 8.5. The table shows that instantaneous detection, i.e. $k_{det} = k_{err}$ is possible within the range of $f_o = 0.35\text{cm}$ and $f_o = 5\text{cm}$. No result shows the CMP effect which means that they are of good quality and provide reliable results using a zonotopic approximation of the united solution set.
The calculation time of all results is also given in Tab. 8.5. It can be seen that the measurement data can be processed in less than the genuine signal time, i.e. $T_{calc} < 120\text{s} = T$ holds for all fault intensities.

Table 8.5: Different offset fault amplitudes for $s_{k_{err}} = h_{2,60} = 37.54\text{cm}$

Fault f_o in cm	Time step of Fault k_{err}	Time step of Detection k_{det}	Quality	Calculation Time T_{calc} in s
5.00	60	60	no CMP	97.2
2.00	60	60	no CMP	92.0
1.00	60	60	no CMP	59.3
0.50	60	60	no CMP	60.1
0.35	60	60	no CMP	57.0
0.20	60	not detected	no CMP	25.2

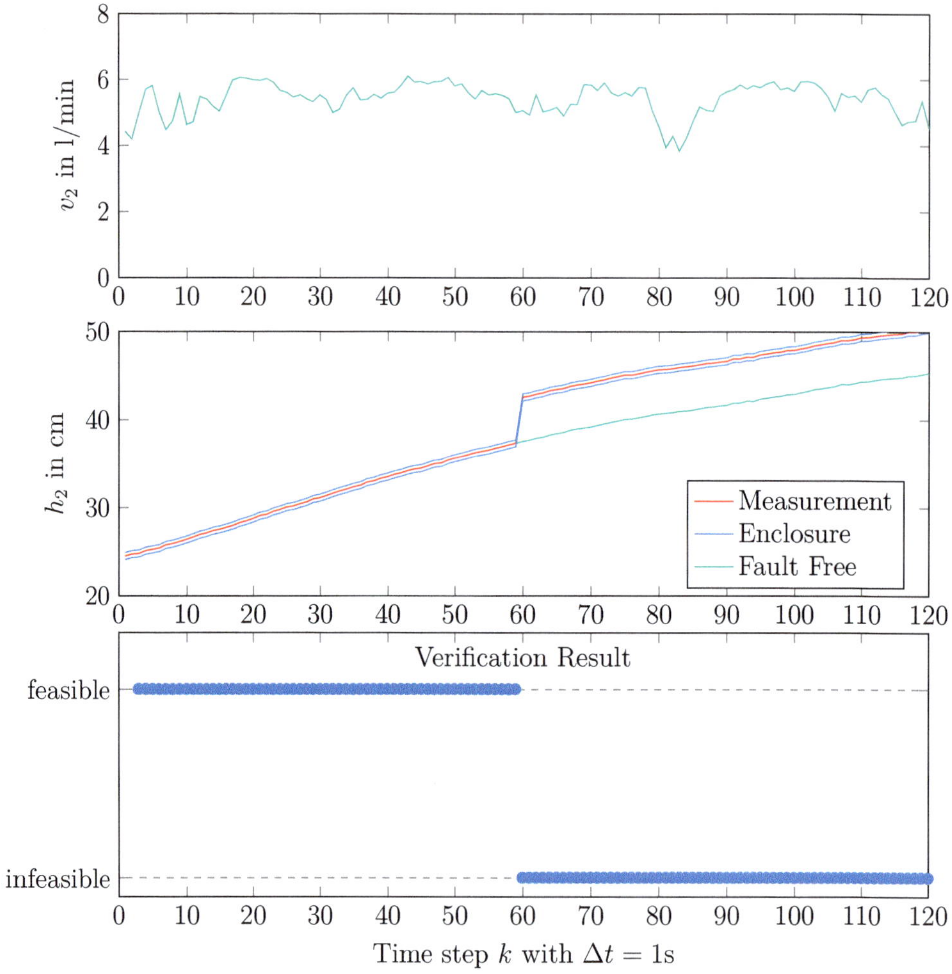

Figure 8.17: Measurement data with offset fault $f_o = 5$cm

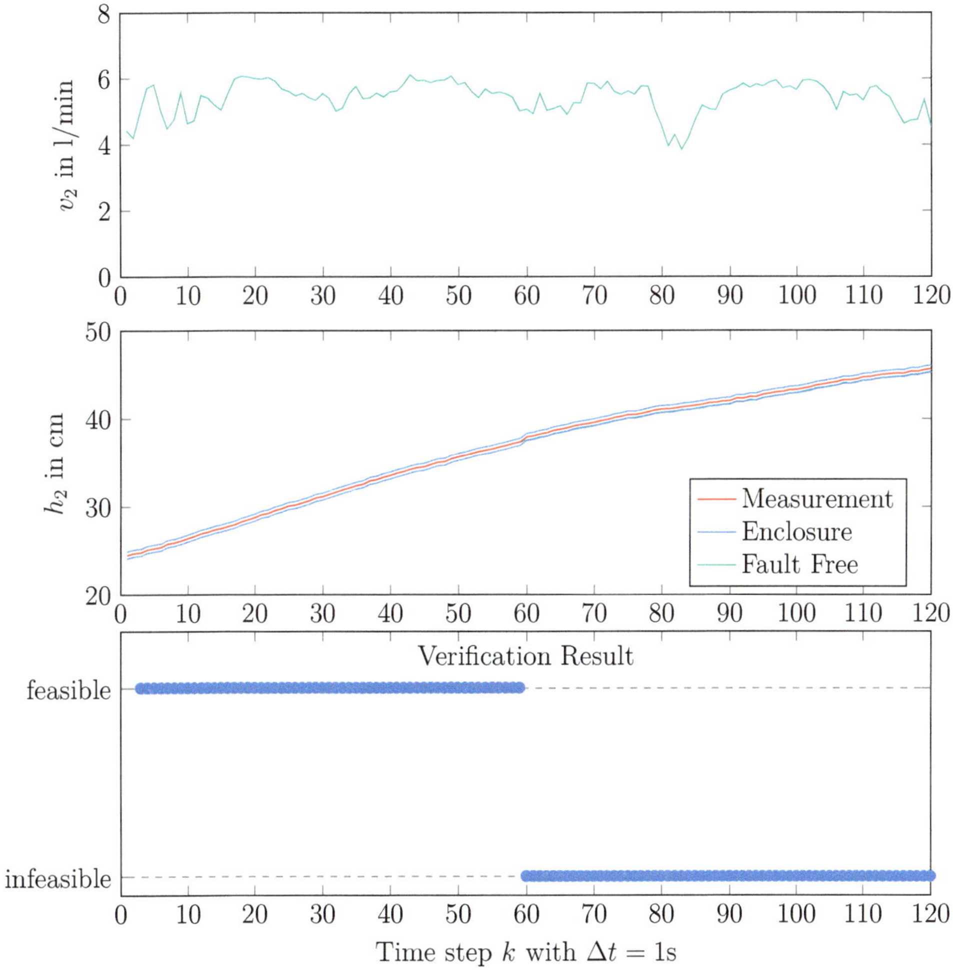

Figure 8.18: Measurement data with offset fault $f_o = 0.35$cm

8.4.3 Scaling Faults

In the diagnosis scenario discussed in this section so far, real measurement data from a single-tank process is used. To realize multiplicative faults with the same measurement data, a scaling fault in the corresponding sensor of h_2 is assumed. This leads to the following scaling fault model

$$s_{s,k} = \begin{cases} s_k & \forall k \in [1, k_{err} - 1] \\ s_k \cdot f_s & \forall k \in [k_{err}, T] \end{cases} \tag{8.43}$$

which replaces the former multiplicative fault (8.42). The absolute deviation to enclose the measurement of h_2 remains $\delta^a_{h_2} = 0.4\text{cm}$. The results are depicted in Fig. 8.19 and Fig. 8.20 for $f_s = 0.95$ and $f_s = 1.01$ respectively.
It can be seen that even factors very close to one (e.g. $f_s = 1.01$, meaning a deviation of 1%) can be detected.

Results for an extensive range of factors are given in Tab. 8.6.
It is not possible to detect the fault intensity of $f_s = 0.97$ as this parameter is very close to the nominal parameter $\boldsymbol{\theta}^* = [0.971,\ 0.979]$ representing the desired system dynamics. This means there is a deviation of 0.1% between $f_s = 0.97$ and $\underline{\theta}^* = 0.971$ which is one order of magnitude less than for $f_s = 1.01$.

All successfully detected faults lead to $k_{det} = 60 = k_{err}$, i.e. they are detected right at their appearance. There was no CMP condition present in the regarded settings.
Again, it is possible to calculate the results for all fault amplitudes in less than the genuine signal time, i.e. $T_{calc} < 120\text{s} = T$.

Table 8.6: Different scaling fault amplitudes for $s_{k_{err}} = h_{2,60} = 37.54\text{cm}$

Fault f_s	Time step of Fault k_{err}	Time step of Detection k_{det}	Quality	Calculation Time T_{calc} in s
1.10	60	60	no CMP	77.7
1.05	60	60	no CMP	80.4
1.03	60	60	no CMP	77.3
1.01	60	60	no CMP	64.6
0.97	60	not detected	no CMP	31.9
0.95	60	60	no CMP	108.8
0.90	60	60	no CMP	96.6
0.75	60	60	no CMP	75.2
0.50	60	60	no CMP	117.8

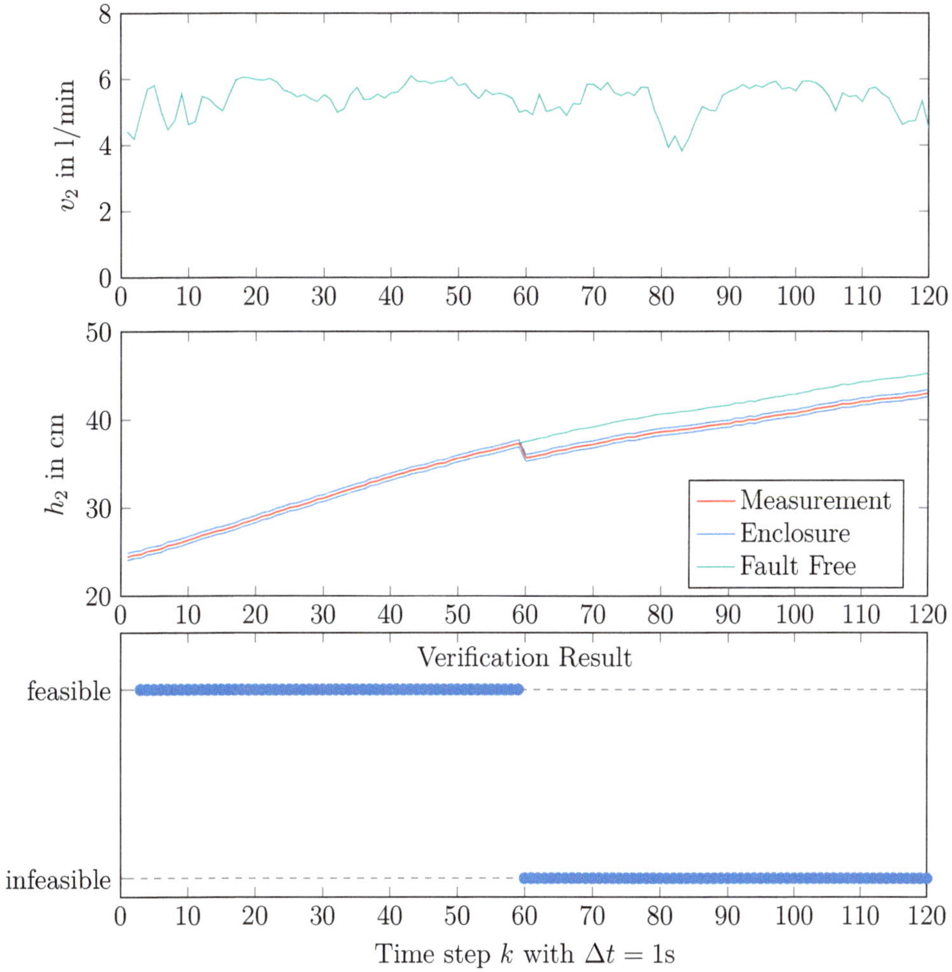

Figure 8.19: Measurement data with scaling fault $f_s = 0.95$

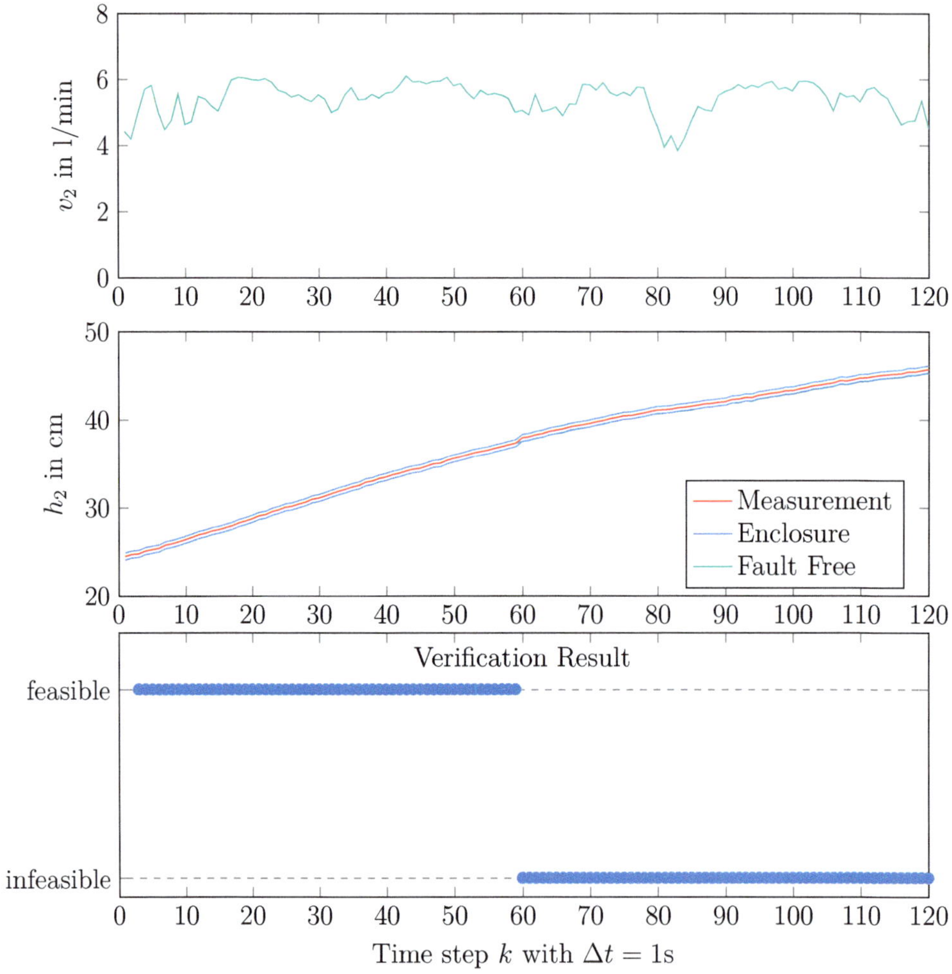

Figure 8.20: Measurement data with scaling fault $f_s = 1.01$

8.5 Conclusion

The methods and theories developed throughout this thesis were applied and demonstrated in this chapter.

First, simulation data of a single-tank process was used to show the performance of the verification method based on Kaucher arithmetic and zonotopic inner enclosures of the united solution set.
This approach was extended to the hybrid setting given by a two-tank system. It was shown that the introduced method is able to verify the correct system in case of known switching times and known active subsystems.
The application to a four-tank process showed that it is not possible to verify the system in various faulty settings even for very small fault amplitudes. This is a relevant indicator that a fault is present in the system and can thus be used for fault detection. The performance of the developed method was shown for three different and common fault types, namely freeze fault, offset fault and multiplicative fault.
The same diagnosis algorithm was finally applied to real measurement data provided again by the single-tank process. It could be shown that the algorithm obtains valuable results by detecting even very small faults from real world data.

The calculation time of all introduced examples was less than the genuine time of experiment on a standard laptop. Therefore the application might in general be suitable for online application in a diagnosis setting.

9 Conclusion

Modern engineering is able to develop and build complex and powerful systems to an unprecedented extend. The functionality of these systems is rapidly increasing and masters tasks that used to be subject to highly trained humans. The challenge how to build such systems is nearly completed. Still remaining is the question how to ensure correct functionality of such powerful safety critical systems. Current safety analysis relies on sophisticated methods from the field of testing. Even though these methods are very mature, they are essentially falsification approaches meaning that there are type II errors by definition. However, in the case of safety critical systems, it is necessary to ensure the absence of type II errors.
This thesis provides the foundations for a new specification and verification approach able to provide the necessary type II error free results.

Therefore a new notion of set based consistency for dynamic systems with a given specification is presented. Kaucher interval arithmetic is used to enclose the measurement data in a bounded error sense. Thus, the specified behavior of a dynamic system can be verified by measurement data even in the presence of noise and sensor uncertainty. Consistency is defined using the Kaucher arithmetic united solution set which leads to mathematically guaranteed results. The verdicts calculated by the new Kaucher based method can not show type II errors (hidden faults) by definition and are thus suitable to provide a reliable verification of safety critical systems.

It was proven mathematically that this holds for a wide class of systems, including time invariant, interval type and hybrid systems, which can be used to describe even nonlinearities. The notion of consistency was extended to include the discrete event part of a hybrid system and requirements on the connection of the two system classes were derived. Several extensions were introduced, leading to a new iterative identification and segmentation algorithm for hybrid systems which is able to handle even unknown switching times. In case the calculations can be done fast enough, the developed approach can also be used for the diagnosis of dynamic systems. Requirements on sampling time and hardware performance have to be determined for each specific setting individually.

The presented methods were successfully applied to several example systems, consisting of a variation of different tank settings. The results were shown, interpreted and discussed.

The results provide the base to answer the research question that governed this thesis. The new theories, methods and algorithms developed in this thesis form the foundation for reliable safety analysis of highly automated safety critical systems. The results of this thesis can be used to solve the arising problems of current powerful and interconnected systems that are increasingly interleaving our daily live.

A Analysis Perspectives

A popular definition to distinguish different analysis perspectives was coined by [Boe84]: validation means "building the right product" whereas verification means "building the product right".

This is used to set up a high level differentiation between *validation* and *verification / falsification* as given in Fig. A.1. Validation is always concerned with the desire of the customer and evaluates the question whether the developed functionality fulfills this desire. The field of validation is very important and large research effort including psychology, behavioral science and linguistics has been put on it in the last decades [Mac95][Que98][Fau03][Fol08][Bar13]. Nevertheless validation is not part of this thesis.

The objective of *security* focuses on the detection of intentional misuse by (un)authorized subjects, e.g. due to hacking attacks or user errors. The whole field of security, including conscious misuse, hacking or manipulation is also not in the scope of this thesis.

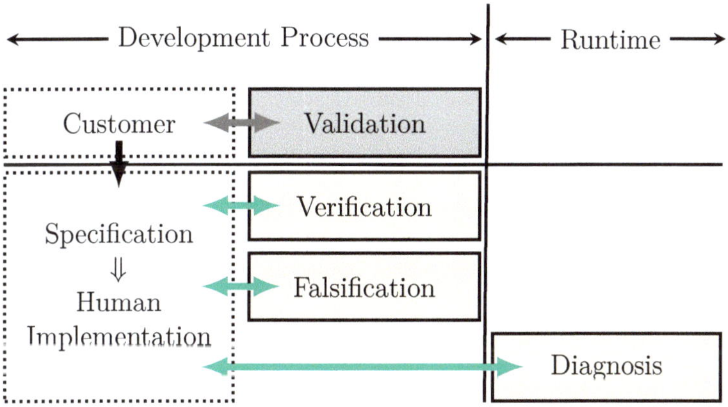

Figure A.1: Evaluation terminology

The realization of the specification is done by human engineers, therefore it is likely that there are mistakes during the process of implementation. A wide spread approach to find these mistakes is given by the concept of falsification, which tries to determine a so called *counter example* that shows unspecified or wrong behavior. If it is not possible to determine a counter example it is assumed that there are no counter examples at all and thus the system is considered to be correct. However, due to limited runtime of the falsification process it is possible that there are undetected (hidden) faults in the implementation. Therefore type II errors are possible which is an disadvantage in case of safety critical systems.

If the specification is very formal, the implemented system can be analyzed in a formal way.

Verification methods aim on proving the correctness of the implemented system in (all) operating conditions. The goal is to prove that the system always shows nominal behavior.
Verification and falsification methods are in general conducted during the development process, while the system operates in some kind of artificial environment. The evaluation can be done offline and might thus need more calculation time or can be run several times during the development process. Mistakes occurring during system operation are tackled by methods of the diagnosis and monitoring field. They need to run online in parallel to the real system operation and are thus required to be very fast.
In case of model based diagnosis, a model of the nominal system is generated that is used to calculate the nominal system output in parallel to the real verification object (VO). Therefore the inputs of the VO are measured and also applied to the nominal model. The resulting outputs are compared with the measured outputs of the VO which leads to a so called residual vector (see among others [Ise93][Ven03a][Ven03c][Ble10]). In case of an undisturbed system, the residual vector is zero if there is no fault present in the VO. A fault is detected if there is a non-zero residual vector. If it is necessary to gain further knowledge of the fault, more sophisticated methods can be applied to localize the exact point of fault occurrence within the VO ([Ven03a][Ven03c][Che14]). A drawback of the diagnosis approach is that - due to measurement noise and model imprecision - the residual vector is not always exactly zero even in the fault free case [Ise06, p. 198].

B Derivation of the Interval Distribution

This appendix provides the derivation of a probability distribution on the parameter p connecting two intervals u and y.
Two examples are presented in Chapter 3. Example 3.4 shows a setting with a proper result of p and Example 3.5 demonstrates a setup leading to an improper solution. The interval ranges of u and y are sampled with $\Delta u = \Delta y = 0.0001$ and used to calculate the resulting parameter p_s for all possible combinations.
It is also possible to theoretically derive the shown results. Therefore, two random variables u and y are defined with uniform distribution between the infimum and the supremum of the interval values $\boldsymbol{u}$ and $\boldsymbol{y}$. The probability density functions of the two random variables are given by:

$$f_u\left(u\right) = \begin{cases} \frac{1}{\overline{u}-\underline{u}} & , \forall u \mid \underline{u} \leq u \leq \overline{u} \\ 0 & , \text{else} \end{cases} \tag{B.1}$$

and

$$f_y\left(y\right) = \begin{cases} \frac{1}{\overline{y}-\underline{y}} & , \forall u \mid \underline{u} \leq u \leq \overline{u} \\ 0 & , \text{else.} \end{cases} \tag{B.2}$$

A general probability density function according to [Bro08, p. 816] has to fulfill the assumptions

$$f\left(x\right) \geq 0,\ \forall x \tag{B.3}$$

$$\int_{-\infty}^{\infty} f\left(x\right) dx = 1. \tag{B.4}$$

Assumption (B.3) is valid for (B.1) and (B.2) by definition. Assumption (B.4) can be shown as follows:

$$\begin{aligned} \int_{-\infty}^{\infty} f_u\left(u\right) du = 1 \\ &= \underbrace{\int_{-\infty}^{\underline{u}} f_u\left(u\right) du}_{0} + \underbrace{\int_{\underline{u}}^{\overline{u}} f_u\left(u\right) du}_{\left[\frac{1}{\overline{u}-\underline{u}} u\right]_{\underline{u}}^{\overline{u}}} + \underbrace{\int_{\overline{u}}^{\infty} f_u\left(u\right) du}_{0} \\ &= \left[\frac{1}{\overline{u}-\underline{u}} \overline{u}\right] - \left[\frac{1}{\overline{u}-\underline{u}} \underline{u}\right] = 1. \end{aligned}$$

□

The proportional parameter p with

$$u \cdot p = y \tag{B.5}$$

thus can be interpreted as random variable

$$p = g(u, y) = \frac{y}{u}. \tag{B.6}$$

There are different possible realizations of u and y depending on the specific values of p. Therefore the probability density function of p is according to [Jon02, p. 118] given as

$$f_p(p) = \int_{-\infty}^{\infty} |u| \, f_u(u) \, f_y(u \cdot p) \, du. \tag{B.7}$$

With the constant densities of the uniform distributions $f_u(u)$ and $f_y(y)$, and assuming only non-negative input values $u > 0$, (B.7) can be relaxed to

$$f_p(p) = \int_{0}^{\infty} u \underbrace{f_u(u)}_{\substack{\text{constant for} \\ \underline{u} \leq u \leq \overline{u} \\ \text{else } 0}} \underbrace{f_y(u \cdot p)}_{\substack{\text{constant for} \\ \underline{y}/u \leq p \leq \overline{y}/u \\ \text{else } 0}} du. \tag{B.8}$$

The antiderivative is zero for all $p \notin [\underline{y}/u, \ \overline{y}/u]$ and $1/p \notin [\underline{u}/y, \ \overline{u}/y]$. Else the densities consist of constant values, leading to the antiderivative

$$f_p(p) = \left[\frac{1}{2} c_u c_y u^2 \right]_0^{\infty} \tag{B.9}$$

with $c_u = \frac{1}{\overline{u} - \underline{u}}$ and $c_y = \frac{1}{\overline{y} - \underline{y}}$. The evaluation of $f_p(p)$ depends on the infimum and supremum of $\boldsymbol{u}$ and $\boldsymbol{y}$ and can be generalized as

$$f_p(p) = \frac{1}{2} c_u c_y \max\left(0, \left(\min\left(\overline{u}, \max\left(\underline{y}/p, \overline{y}/p\right)\right)\right)^2 - \left(\max\left(\underline{u}, \min\left(\underline{y}/p, \overline{y}/p\right)\right)\right)^2\right). \tag{B.10}$$

It is now possible to draw the derived density function for specific values of $\boldsymbol{u}$ and $\boldsymbol{y}$. Exemplary plots for a proper and an improper setting are given in Example B.1.

Example B.1:

The plot for the values of $\boldsymbol{u} = [2,\ 3]$ *and* $\boldsymbol{y} = [4,\ 9]$ *according to Example 3.4 is depicted in Fig. B.1. The density function has the same shape as the sampling based result given in Fig. 3.5. The plateau in the figure shows that the solution is proper.*

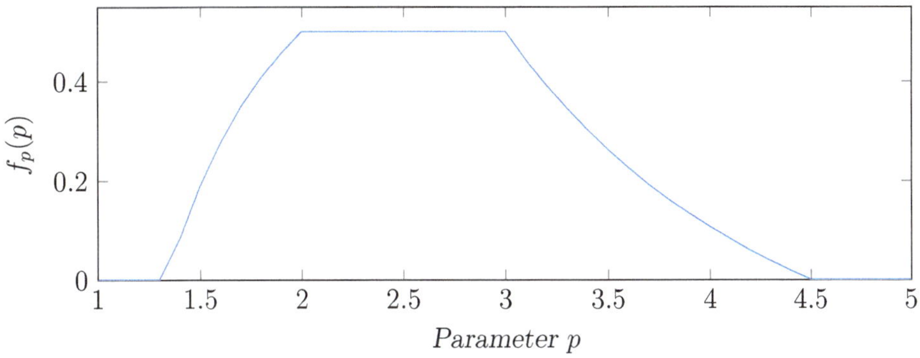

Figure B.1: Probability density function $f_p(p)$ for the proper case

The plot of an improper setting according to Example 3.5 is given in Fig. B.2. The input and output intervals are $\boldsymbol{u} = [2,\ 3]$ *and* $\boldsymbol{y} = [4,\ 5]$ *leading to an improper solution and an eroded plateau.*

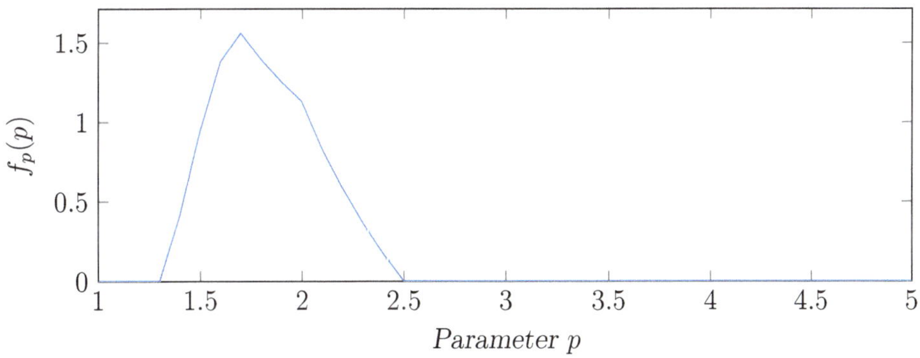

Figure B.2: Probability density function $f_p(p)$ for the improper case

These results are in accordance with the theory introduced in Chapter 3.

C Full Rank Criteria

It is in general NP-hard to determine whether a given interval matrix $\boldsymbol{A}$ has full rank, respectively to check the matrix for singularity [Sha14]. However, there are some criteria to determine the property of full rank [Sha14]. Four sufficient conditions and one necessary and sufficient condition are given in the following.
It is necessary to introduce the absolute value of an interval

$$|\boldsymbol{x}| = \max\left(|\underline{x}|, |\overline{x}|\right) \tag{C.1}$$

and the magnitude

$$x^{+} = \begin{cases} \min\left(|\underline{x}|, |\overline{x}|\right) & \text{, if } 0 \notin \boldsymbol{x} \\ 0 & \text{, else.} \end{cases} \tag{C.2}$$

The first sufficient condition for quadratic problems is based on diagonal dominance. The interval matrix $\boldsymbol{A} \in \mathbb{IR}^{(n \times n)}$ is nonsingular, if it is diagonal dominant. This means the inequality

$$a^{(ii)+} > \sum_{\substack{j=1 \\ j \neq i}}^{n} \left|\boldsymbol{a}^{(ij)}\right| \tag{C.3}$$

holds for $i \in \{1, 2, \ldots, n\}$.
There are two approaches to extend this condition to overdetermined equation systems, i.e. $\boldsymbol{A} \in \mathbb{IR}^{(m \times n)}$ with $m > n$. The first approach searches for diagonal dominant subsquares within the overdetermined interval matrix. If there is such a diagonal dominant subsquare, the whole interval matrix has full rank. However there might be no diagonal dominant subsquare even though the matrix has full rank. This can be due to permutation of rows of the matrix. Even though permuted lines do not change the rank of a matrix, it does change the appearance of diagonal dominant subsquares. However, the property that permuting rows does not change the rank of the matrix can also be used to solve the problem. The lines can be permuted algorithmically such that diagonal dominant subsquares are created. This condition is still sufficient for overdetermined systems.

A second sufficient condition for full rank of an interval matrix $\boldsymbol{A} \in \mathbb{IR}^{(m\times n)}$ is based on the spectral radius. Thereby the spectral radius $\rho(A)$ is defined to be the largest absolute singular value of the matrix A [Lax02, p. 195]. If the spectral radius fulfills

$$\rho\left(\left|(A_c)^\dagger\right| A_\Delta\right) < 1 \tag{C.4}$$

and the center matrix A_c has full rank, also the interval matrix $\boldsymbol{A}$ has full rank. Thereby $A^\dagger = \left(A^T A\right)^{-1} A^T$ denotes the pseudo inverse of the matrix $A \in \mathbb{R}^{(m\times n)}$ with $m \geq n$.

A third sufficient condition is based on the singular values of the matrix $\boldsymbol{A} \in \mathbb{IR}^{(m\times n)}$. If the condition

$$\sigma_{\max}(A_\Delta) < \sigma_{\min}(A_c) \tag{C.5}$$

is fulfilled, the interval matrix $\boldsymbol{A}$ has full rank. Thereby $\sigma_{\max}(A)$ and $\sigma_{\min}(A)$ denote the greatest, respectively smallest, singular value of the matrix A. The singular values are defined as the nonnegative solutions to the system

$$\begin{pmatrix} 0 & A^T \\ A & 0 \end{pmatrix} \begin{pmatrix} x \\ y \end{pmatrix} = \sigma \begin{pmatrix} x \\ y \end{pmatrix}. \tag{C.6}$$

Condition four uses an absolute subordinate matrix norm $||\cdot||$ of $\boldsymbol{A} \in \mathbb{IR}^{(n\times m)}$. Assuming full rank of the center matrix A_c, the sufficient condition is given by

$$||A_\Delta|| < ||A_c^\dagger||^{-1}. \tag{C.7}$$

If (C.7) holds, $\boldsymbol{A}$ has full rank. The proofs of all four sufficient conditions are given in [Sha14].

According to [Roh12] there is a fifth, necessary and sufficient condition: An interval matrix $\boldsymbol{A} \in \mathbb{IR}^{(m\times n)}$ with $m \geq n$ has full rank iff

$$|A_c X| \leq A_\Delta |X| \tag{C.8}$$

with $X \in \mathbb{R}^n$ can only be solved by the zero solution $X = [0, 0, \ldots, 0]^T$.
Sufficiency is based on the idea that there is a non trivial solution $X \neq [0, 0, \ldots, 0]^T$ as soon as the matrix $\boldsymbol{A}$ does not have full rank. Necessity follows from the existence of the non trivial solution. If this is the case, $\boldsymbol{A}$ cannot have full rank or the non trivial solution does not solve (C.8). An extensive proof of this condition is given in [Sha14]. However, the approach directly aims on an NP-Hard problem which means that it can only be checked approximately.

D Existence and Uniqueness of the Algebraic Solution Set

The sufficient conditions on the existence of an algebraic solution given in [Sha96], [Mar99] and [Lak99] are sketched in this appendix. To follow those ideas, two more interval arithmetic notations are necessary.
Using the dual $(\cdot)$ operator given in (3.37), the proper projection pro $(\boldsymbol{x})$ is defined as

$$\text{pro}\,(\boldsymbol{x}) = \begin{cases} \boldsymbol{x} & \text{, if } \boldsymbol{x} \text{ is proper} \\ \text{dual}\,(\boldsymbol{x}) & \text{, else.} \end{cases} \tag{D.1}$$

The second property is $\imath$-nonsingularity. A quadratic point real matrix $Q \in \mathbb{R}^{(n\times n)}$ is called $\imath$-nonsingular if

$$Qx = 0 \Leftrightarrow x = 0 \in \mathbb{IR}^n \tag{D.2}$$

holds. Otherwise Q is called $\imath$-singular.

According to [Sha96] there is an algebraic solution $\sum_a$ to the interval linear equation $\boldsymbol{A}x = \boldsymbol{B}$ with $\boldsymbol{A} \in \mathbb{IR}^{(n\times n)}$ for any $\boldsymbol{B} \in \mathbb{IR}^n$, if $\boldsymbol{A}$ is sufficiently narrow and pro $(\boldsymbol{A})$ contains an $\imath$-nonsingular point matrix.

Thereby "sufficiently narrow" means that $|A_\Delta|$ is sufficiently small.

The proof of existence given in [Mar99] is based on the iterative approach to determine the algebraic solution set given in [Kup95]. For this proof, the notation of the diagonal matrix $D(\boldsymbol{A})$ is introduced for an interval matrix $\boldsymbol{A} \in \mathbb{IR}^{(n\times n)}$

$$D(\boldsymbol{A}) = \left(\boldsymbol{d}^{(i,j)}\right)_{1\le i\le n, 1\le j\le n} = \begin{cases} \boldsymbol{a}^{(i,j)} & \text{, if } i = j \\ 0 & \text{, if } i \neq j. \end{cases} \tag{D.3}$$

The inverse of the diagonal matrix is given by

$$D^{-1}(\boldsymbol{A}) = \left(\tilde{\boldsymbol{d}}^{(i,j)}\right)_{1\le i\le n, 1\le j\le n} = \begin{cases} 1/\text{dual}\left(\boldsymbol{a}^{(i,j)}\right) & \text{, if } i = j \\ 0 & \text{, if } i \neq j. \end{cases} \tag{D.4}$$

Using the dual $(\cdot)$ operator from (3.37). The iterative solution algorithm given in [Kup95] converges to the algebraic solution $\sum_a$ if

$$||D^{-1}(\boldsymbol{A})|| \leq 1 \tag{D.5}$$

$$||\boldsymbol{A} + \text{opp}\,(D(\boldsymbol{A}))|| \leq 1 \tag{D.6}$$

holds, with opp $(\cdot)$ according to (3.34). The used matrix norm $||\cdot||$ is the maximum of the linewise sum of the absolute interval values (C.1):

$$||\boldsymbol{A}|| = \max_{1 \leq i \leq n} \left(\sum_{k=1}^{n} \left| \boldsymbol{a}^{(i,k)} \right| \right). \tag{D.7}$$

The interested reader is referred to [Mar99] for further considerations.

A generalized approach for overdetermined systems $\boldsymbol{A} \in \mathbb{IR}^{(m \times n)}$ was introduced in [Lak99]. The regressor matrix $\boldsymbol{A}$ is split in three parts with $\boldsymbol{A} = \boldsymbol{A}_0 + \boldsymbol{A}_1 + \boldsymbol{A}_2$ and

$$\boldsymbol{A}_0 = \left(\boldsymbol{a}_0^{(i,j)}\right)_{1 \leq i \leq m, 1 \leq j \leq n} = \begin{cases} \boldsymbol{a}^{(i,j)} & \text{, if } \underline{a}^{(i,j)}\overline{a}^{(i,j)} \geq 0 \\ 0 & \text{, else} \end{cases} \tag{D.8}$$

$$\boldsymbol{A}_1 = \left(\boldsymbol{a}_1^{(i,j)}\right)_{1 \leq i \leq m, 1 \leq j \leq n} = \begin{cases} \boldsymbol{a}^{(i,j)} & \text{, if } \underline{a}^{(i,j)} < 0 < \overline{a}^{(i,j)} \\ 0 & \text{, else} \end{cases} \tag{D.9}$$

$$\boldsymbol{A}_2 = \left(\boldsymbol{a}_2^{(i,j)}\right)_{1 \leq i \leq m, 1 \leq j \leq n} = \begin{cases} \boldsymbol{a}^{(i,j)} & \text{, if } \underline{a}^{(i,j)} > 0 > \overline{a}^{(i,j)} \\ 0 & \text{, else.} \end{cases} \tag{D.10}$$

The problem can be reformulated as an extended system that considers the upper and lower bounds of the interval values explicitly, as given in [Lak99]. This problem can then be transferred to a set of inequality conditions. It is possible to show that there is not more than one solution for any $b \in \boldsymbol{B}$ if the derived set of inequality conditions has zero as unique solution. The extensive proof is given in [Lak99].

E System Behavior Specification

The verification methods developed in this thesis are based on the assumption of a system specification available in ARX form. In a practical setting, it is necessary to determine these nominal parameters. Control engineering specifications are in general based on tolerance bands, steady-state errors, rise and settling times or acceptable overshoots. Such specifications inherently show interval properties - even though in general no interval arithmetic is used.

This appendix provides two approaches to determine the ARX parameters of such intuitive graphic specifications. In the time domain, a method is introduced to determine the parameters from a desired step response. This step response can be set up using the drag-and-drop function provided by a toolbox. A second method determines the parameters from the frequency domain. Therefore only the tolerance band widths and pass/cut-off frequencies of a filter function need to be specified. In case the Kaucher based method is applied in a diagnosis setting, it is beneficial to use a nominal physical model of the regarded process. This physical model can also be used to determine the desired ARX parameters.

E.1 Time Domain Specification

This specification approach allows an intuitive specification of the desired behavior based on time domain input-output behavior. The first step is to specify an input signal and the desired resulting output signal. Also the class of the desired system behavior has to be given. The time domain input-output behavior is then used to determine the parameters of a transfer function in the complex s-plane. In this appendix a step input is used to determine the properties of a proportional gain first order time delay system (PT1). The method given in [Föl13, p. 77] can be used to determine the system parameters based on a given step response. A time continuous step response denotes the output values $y(t)$ generated by a step input

$$u(t) = \sigma(t) = \begin{cases} 0, & k < 0 \\ 1, & k \geq 0. \end{cases} \tag{E.1}$$

The complex frequency domain transfer function of a basic PT1 system is given by

$$G(s) = \frac{k_p}{1 + \tilde{T}s} \tag{E.2}$$

with gain k_p and time delay $\tilde{T}$. If both parameters are determined, $G(s)$ can be transformed to its discrete time representation. The desired nominal parameters Θ^* are then given by the parameters of the discrete time transfer function.

For given input-output data $[u(t), y(t)]$ with $t \in [0,\ T]$ it is possible to determine the transfer function parameters. The gain k_p is given by the stationary value $k_p = y_\infty$ which is defined to be the last point of the measurement $y(T)$, assuming T is large enough to allow $y(t)$ to settle. If the output signal is sampled with sampling time Δt, the information of the resulting points can be used to calculate the time delay $\tilde{T}$. Therefore each available sampling point $y_k = y(k\Delta t)$ is used to calculate an auxiliary value

$$\eta_k = 1 - \frac{y_k}{y_\infty} \tag{E.3}$$

which is then used to determine the time delay

$$\tilde{T}_k = -\frac{k\Delta t}{\ln(\eta_k)}. \tag{E.4}$$

The time delay of the transfer function $\tilde{T}$ is given by the arithmetic mean of all $n_k = \frac{T}{\Delta t}$ values of $\tilde{T}_k$, i.e.

$$\tilde{T} = \frac{1}{n_k} \sum_{k=1}^{n_k} T_k. \tag{E.5}$$

Afterwards the transformation to discrete time is done. The resulting parameters Θ^* are used to determine the ARX step response.
The introduced functionality is implemented in a Toolbox. The application of this Toolbox is demonstrated in Example E.1.

Example E.1:

Based on an initial arbitrary input-output signal, the toolbox provides the possibility to move the sampling points via drag-and-drop. Fig. E.1 shows an exemplary output signal that was created as the step response of a PT1 system with gain $k_P = 10$ *and* $\tilde{T} = 2$. *The resulting continuous trajectory was sampled with* $\Delta t = 1\text{s}$. *The blue points depict the sampling points that can be moved using drag-and-drop. The user can move the sampling points such that the resulting trajectory shows the desired behavior. In this case, the toolbox calculates the system parameters according to (E.3)-(E.5). The resulting system is able to generate the desired values for the given step input. The depicted setting leads to the time discrete ARX parameters* $a = 0.64227, c = 3.6077$. *The respective time discrete step response is given by the red trajectory in Fig. E.1.*

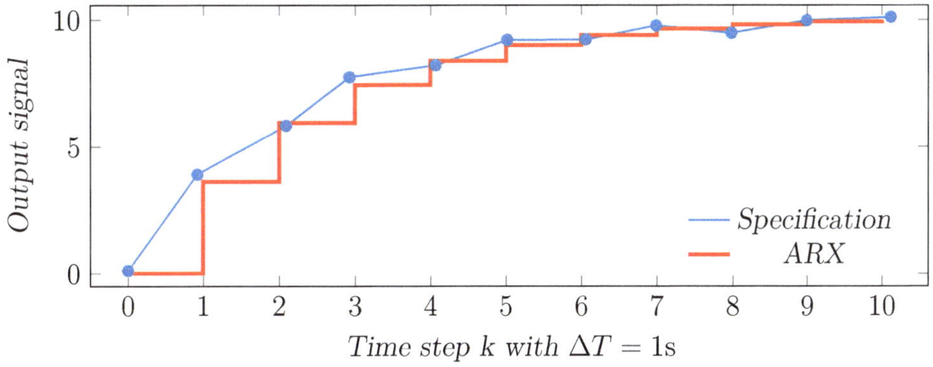

Figure E.1: Time domain specification toolbox

It can be seen in Fig. E.1 that the time discrete trajectory (red) is close to the specified points (blue). This demonstrates that it is possible to determine the ARX parameters of a system by using a graphical user interface with drag-and-drop to set up the desired step response.

E.2 Frequency Domain Specification

This specification approach allows an intuitive specification of the desired behavior based on designing the amplitude response of the system. First order systems can be interpreted as low pass filters. Each Filter has a specific frequency domain characteristic consisting of the location and the width of the passband and the stopband. Based on this information, the method of [Lüc80, p. 147ff] is used to determine the filter coefficients. The resulting filter can be transformed to discrete time which leads to the desired ARX coefficients.

The method is applicable for low-pass, high-pass, bandpass and band-rejection filters. In this appendix the design of a low-pass is presented. Therefore the cut off frequency Ω_p, the stop band frequency Ω_s and the respective passband width Δ_p and stopband width Δ_s (see Fig. E.2) need to be defined by the user. These frequencies are defined with respect to the periodic interval of the frequency response $0 \leq \Omega \leq \pi$. The frequencies are now transformed into the $0 \leq f \leq \infty$ domain by

$$f_p = \tan\left({}^{\Omega_p}/_2\right) \tag{E.6}$$
$$f_s = \tan\left({}^{\Omega_s}/_2\right). \tag{E.7}$$

These values lead to the normalized low-pass representation

$$f_{p,norm} = 1 \tag{E.8}$$
$$f_{s,norm} = {}^{f_s}/_{f_p}. \tag{E.9}$$

The normalized low-pass representation can be achieved for all four kinds of filters, by using different transformations. The following design routine is thus applicable in every setting. To ensure that the given passband and stopband limits are met, the auxiliary variables

$$\tilde{\Delta}_d = \frac{\sqrt{2\Delta_d - \Delta_d^2}}{1 - \Delta_d}, \text{ for } 0 \leq f \leq 1 \text{ (Passband)} \tag{E.10}$$
$$\tilde{\Delta}_s = \frac{\sqrt{1 - \Delta_s^2}}{\Delta_s}, \quad \text{for } f_{s,norm} \leq f \text{ (Stopband)} \tag{E.11}$$

are calculated. The transfer function of an exponential filter is then given by

$$G(f) = G_0 \frac{1}{\prod_{i=1}^{q} f - f_{\infty,i}} \tag{E.12}$$

with poles $f_{\infty,i}$ and normalization constant G_0. The denominator order $q \in \mathbb{N}$ can be calculated with

$$q \geq \frac{\log_{10}\left({}^{\tilde{\Delta}_s}/_{\tilde{\Delta}_d}\right)}{\log_{10}\left(f_{s,norm}\right)}. \tag{E.13}$$

The real part and the imaginary part of a complex pole $f_{\infty,i}$ are given by

$$\Re\left(f_{\infty,i}\right) = -\epsilon^{-1/q} \sin\left(\frac{2i-1}{q}\frac{\pi}{2}\right) \tag{E.14}$$
$$\Im\left(f_{\infty,i}\right) = \epsilon^{-1/q} \cos\left(\frac{2i-1}{q}\frac{\pi}{2}\right). \tag{E.15}$$

The factor ϵ and the respective normalization can be used to adjust the level of the frequency response. The used value of ϵ can be chosen from the interval

$$\epsilon = [\underline{\epsilon}, \overline{\epsilon}] = \left[\frac{\tilde{\Delta}_s}{\left(f_{s,norm}\right)^q}, \tilde{\Delta}_d\right]. \tag{E.16}$$

Thereby choosing $\epsilon = \overline{\epsilon}$ means that the frequency response touches the constrained region at the end of the passband $f_{p,norm}$, whereas $\epsilon = \underline{\epsilon}$ means that it touches the constrained region at the beginning of the stopband $f_{s,norm}$.
The normed lowpass is now transformed back to its genuine frequency form and afterwards into the time discrete representation in order to extract the desired nominal parameters Θ^* of the transfer function. An application of this method is given in Example E.2.

Example E.2:
The frequency domain specification of a low-pass filter is given by $\Delta_p = 0.1$, $\Delta_s = 0.2$, $\Omega_p = 0.4\pi$ *and* $\Omega_s = 0.7\pi$. *This means that the desired frequency response is located within the green area in Fig. E.2. The introduced filter design procedure of [Lüc80, p. 147ff] is applied to the setting.*
Based on the specified values, the choice $\epsilon = \overline{\epsilon}$ *leads to the ARX coefficients*

$$[a_1, a_2, a_3] = [0.1425, -0.3387, 0.0130] \tag{E.17}$$
$$[c_1, c_2, c_3, c_4] = [0.1479, 0.4437, 0.4437, 0.1479]\,. \tag{E.18}$$

The respective frequency response is depicted as solid blue line in Fig. E.2.
Using the same values but choosing $\epsilon = \underline{\epsilon}$ *leads to*

$$[a_1, a_2, a_3] = [-0.2643, -0.3518, -0.0244] \tag{E.19}$$
$$[c_1, c_2, c_3, c_4] = [0.2051, 0.6152, 0.6152, 0.2051]\,, \tag{E.20}$$

displayed as dashed line in Fig. E.2.

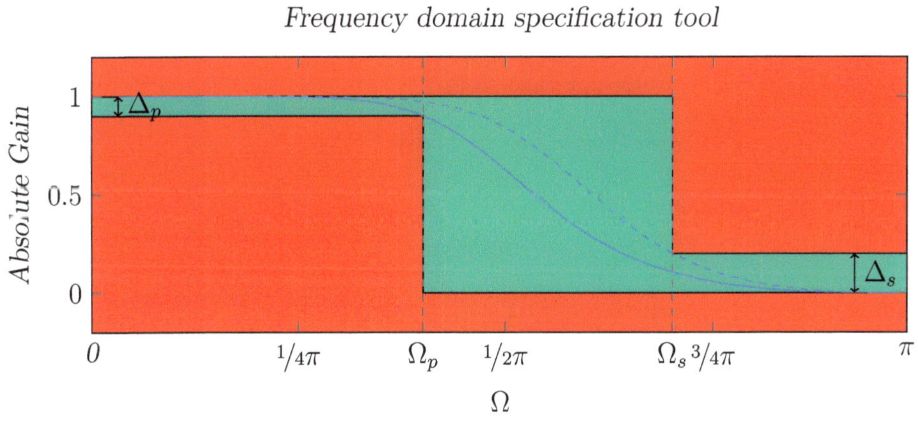

Figure E.2: Frequency domain specification toolbox

F Excitation Signal Design

Hybrid system verification poses specific requirements on the excitation signal. It is assumed that these requirements are fulfilled throughout this thesis. However, determining a suitable excitation signal is in general not trivial. The excitation signal of a measurement is often chosen depending on the intended purpose of the experiment. Arbitrary noise signals (white Gaussian noise) can be used to ensure persistent excitation of all frequencies. Arbitrary meaningful signals (impulse or step signal) are used to perform control theoretic modeling such as impulse response or step response. Also there are specifically designed input signals fitted to the implemented logic in the current verification object.
In the context of hybrid systems as regarded in this thesis, there are two properties that need to be fulfilled. Each subsystem with its respective individual dynamic needs to be persistently excited. Furthermore, all states of the superimposed state machine need to be activated once. Therefore the respective switching thresholds have to be met to enable the switch event. The situation that a switch is triggered during the excitation and identification phase of each subsystem has to be avoided. A first possible solution idea was developed in the master thesis [Rie17]. The basic outline is sketched in this appendix. The method uses three steps:

1. Path calculations to ensure state coverage of the superimposed state machine
2. Design of a persistent excitation signal without leaving the subsystem
3. Efficient transfer of the subsystem to its switch threshold

F.1 Path Calculation

The superimposed state machine is transformed to its graph representation $\mathcal{G}_Z$. A coverage algorithm is used to determine paths that include all states and all transitions of the graph. Additionally, the length of the path needs to be minimal to enable short measurement times. If such an optimal excitation signal is used, missing states or transitions can be used to prove inconsistency.
The problem of state and transition coverage can be reduced to transition coverage only. This is due to the structure of the specification, where each state needs to be connected to a transition.

To use a modified depth first state coverage algorithm, the graph $\mathcal{G}_Z$ is transformed such that all transitions are represented by states in the transformed graph $\mathcal{G}'_Z$ and vice versa (see Fig. F.1).

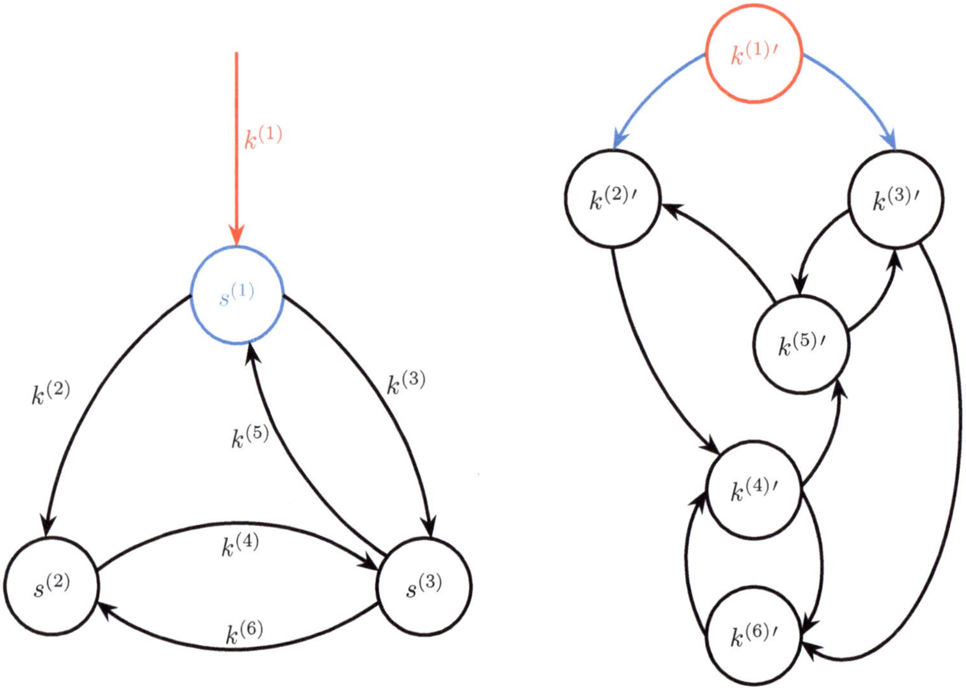

Figure F.1: Genuine graph $\mathcal{G}_Z$ (left) and transformed graph $\mathcal{G}'_Z$ (right)

A modified recursive depth first algorithm is started in a specified initial state and checks the number of possible successors of each successor of the current state. Also the distance to the successors are taken into account to enable short paths. The result of the algorithm is a path that covers all states of $\mathcal{G}'_Z$ and thus all transitions of $\mathcal{G}_Z$. This result is used to identify each subsystem in the path.

F.2 Persistent Excitation Based on Fisher Information Matrix

Persistent excitation of each subsystem is ensured by a specific input signal. This input signal is calculated based on the Fisher information matrix [Eba14][Man10], which is only applicable for stable systems. Using the Fisher information matrix M, the parameter covariance of an estimator is limited by the Cramer Rao Bound [Goo77] to

$$cov(\Theta) \geq \frac{1}{M}. \tag{F.1}$$

Thereby M is defined using the expectation $E\{\cdot\}$ as

$$M(\Theta, U) = E\left\{\left[\frac{\partial log\ (p\,(y \mid \Theta, U))}{\partial\Theta}\right]^T \left[\frac{\partial log\ (p\,(y \mid \Theta, U))}{\partial\Theta}\right]\right\}. \tag{F.2}$$

The probability $log\,p(y \mid \Theta, U)$ resembles the situation that y is observed if the true parameters are given by Θ while using the input $U = \langle u_k \rangle_{k=1}^{N}$. To achieve a parameter covariance as close as possible to the Cramer Rao Bound, the Fisher matrix has to be maximal with respect to the input signal used and the parameters. This can be achieved by using D-optimality for the given nominal parameters as defined in [Man10]:

$$U^* = -\min_{U}(log\ det(M(\Theta^*, U))). \tag{F.3}$$

To solve the optimization Problem F.3, an initial feasible input signal U_{init} is chosen. This signal is then optimized iteratively for each time step $u_{init,k}$ until the optimization converges. Each input value u_k^* is thereby bounded to the range of feasible input values given by the user.

F.3 Transfer to the Switch Threshold

After the identification of the subsystem, it is necessary to activate the successive switch. This is done by transferring the relevant system value within its activation limits $l^{(i)}$.
The specific event and thus the activated transition are already determined in the result of the path calculation. This is done using the well known Hamilton formalism. Therefore the objective function is set up in terms of the difference between the desired value $y_{l_c^{(i)}} = \frac{1}{2}\left(\underline{l}^{(i)} + \bar{l}^{(i)}\right)$ and the current value y_k, i.e. $\Delta y_k = y_k - y_{l_c^{(i)}}$. The resulting objective function is given by

$$J = \frac{1}{2}\Delta y_T S \Delta y_T + \frac{1}{2}\sum_{k=1}^{T-1} \Delta y_k Q \Delta y_k \tag{F.4}$$

with the penalty matrices $S = \frac{1}{\epsilon_{des}}$ and $Q = 1$. It is now possible to set up and solve the Hamilton equations [Sag68]. The first step is to transfer the ARX system description to a vector matrix notation:

$$\begin{bmatrix} y_{k+1} \\ y_k \\ \vdots \\ y_{k-n_a+2} \end{bmatrix} = \underbrace{\begin{bmatrix} a_1 & a_2 & \dots & a_{n_a} \\ 1 & 0 & \dots & 0 \\ \vdots & \ddots & \ddots & \vdots \\ 0 & \dots & 1 & 0 \end{bmatrix}}_{A} \underbrace{\begin{bmatrix} y_k \\ y_{k-1} \\ \vdots \\ y_{k-n_a+1} \end{bmatrix}}_{\tilde{Y}_k} + \underbrace{\begin{bmatrix} c_1 & \dots & c_{n_c} \\ 0 & \dots & 0 \\ \vdots & \ddots & \vdots \\ 0 & \dots & 0 \end{bmatrix}}_{C} \underbrace{\begin{bmatrix} u_k \\ u_{k-1} \\ \vdots \\ u_{k-n_c+1} \end{bmatrix}}_{\tilde{U}_k}. \tag{F.5}$$

Then the Hamilton equation is set up:

$$H(y_k, u_k, k) = \frac{1}{2}\Delta y_k S \Delta y + \lambda_{k+1}^T (A\tilde{Y}_k + C\tilde{U}_k) \tag{F.6}$$

The derivative of H is given by

$$\frac{\partial H}{\partial u} = \lambda_k \mathbf{C}^T. \tag{F.7}$$

with

$$\lambda_k = \Delta \mathbf{y}_k + \mathbf{A}^T \lambda_{k+1} \tag{F.8}$$

leading to

$$\Delta u = -\alpha \frac{\partial H}{\partial u}, \tag{F.9}$$

$$u_k^{(i+1)} = u_k^{(i)} + \Delta u_k. \tag{F.10}$$

The parameter α is used to scale the result in case the calculated solution violates the feasible input range.

This procedure is applied to all subsystems within the calculated path to construct the overall excitation signal. This signal is then applied to the VO and the resulting output values are measured. The resulting input and output measurement data can then be used in any of the methods introduced in this thesis.

G Tables of Geometric Parameters

The parameters of the three-tank lab setting at the Institute of Control Systems (IRS) are given in Tab. G.1.

Table G.1: System properties of the IRS three-tank lab setting

	Value	Unit	Property
h_{32u}	30.0	cm	Height of upper connection valve
h_{32l}	0.0	cm	Height of lower connection valve
a_1	0.5	cm^2	Cross section nominal outflow tank 1
a_2	0.5	cm^2	Cross section nominal outflow tank 2
a_3	0.5	cm^2	Cross section nominal outflow tank 3
a_{13u}	0.5	cm^2	Cross section upper connection valve v_{13u}
a_{13l}	0.5	cm^2	Cross section lower connection valve v_{13l}
a_{32u}	0.5	cm^2	Cross section upper connection valve v_{32u}
a_{32l}	0.5	cm^2	Cross section lower connection valve v_{32l}
a_{leak}	0.8	cm^2	Cross section leakage outflow (only tank 2)
A_1	154.0	cm^2	Cross section tank 1
A_2	154.0	cm^2	Cross section tank 2
A_3	154.0	cm^2	Cross section tank 3
g	981.0	cm/s^2	Gravitational force
γ_1	16.7	$\min cm^3/(l\,s)$	Constant tank 1
γ_2	16.7	$\min cm^3/(l\,s)$	Constant tank 2

The parameters of the four-tank simulation setting are given in Tab. G.2.

Table G.2: System properties of the simulated four-tank lab setting

	Value	Unit	Property
v_{in1}	0.7		Valve 1 flow to tank 1
a_1	0.071	cm^2	Cross section nominal outflow tank 1
a_2	0.071	cm^2	Cross section nominal outflow tank 2
a_3	0.071	cm^2	Cross section nominal outflow tank 3
a_4	0.071	cm^2	Cross section nominal outflow tank 4
A_1	28.0	cm^2	Cross section tank 1
A_2	28.0	cm^2	Cross section tank 2
A_3	28.0	cm^2	Cross section tank 3
A_4	28.0	cm^2	Cross section tank 4
g	981.0	$\mathrm{cm/s}^2$	Gravitational force
γ_1	3.33	cm^3/s	Geometric constant
γ_2	3.33	cm^3/s	Geometric constant

References

Public References

[Abr96] Abrial, J.-R. *The B-book: Assigning Programs to Meanings.* Cambridge University Press, 1996.

[AI15] Araiza-Illan, D., Eder, K. and Richards, A. *Verification of Control Systems Implemented in Simulink with Assertion Checks and Theorem Proving: A Case Study.* In *European Control Conference (ECC)*, pages 2670–2675, 2015.

[AI17] Ait-Izem, T., Harkat, M.-F., Djeghaba, M. and Kratz, F. *Sensor Fault Detection Based on Principal Component Analysis for Interval-valued Data.* Quality Engineering, pages 1–13, 2017.

[Alt08] Althoff, M., Stursberg, O. and Buss, M. *Verification of Uncertain Embedded Systems by Computing Reachable Sets based on Zonotopes.* Proceedings of the 17th IFAC World Congress, volume 41, pages 5125–5130, 2008.

[Alu06] Alur, R., Dang, T. and Ivančić, F. *Predicate Abstraction for Reachability Analysis of Hybrid Systems.* ACM Transations on Embedded Computing Systems, volume 5, pages 152–199, 2006.

[Apo67] Apostolatos, N. and Kulisch, U. *Grundlagen einer Maschinenintervallarithmetik.* Computing, volume 2, pages 89–104, 1967.

[Ara17] Araujo, H., Carvalho, G., Sampaio, A., Mousavi, M. R. and Taromirad, M. *A Process for Sound Conformance Testing of Cyber-Physical Systems.* In *IEEE International Conference on Software Testing, Verification and Validation Workshops (ICSTW)*, pages 46–50, 2017.

[Arm09] Armengol, J., Vehi, J., Sainz, M. A., Herrero, P. and Gelso, E. R. *SQualTrack: A Tool for Robust Fault Detection.* IEEE Transactions on Systems, Man, and Cybernetics, volume 39, pages 475–488, 2009.

[Asa06] Asarin, E., Dang, T., Frehse, G., Girard, A., Guernic, C. L. and Maler, O. *Recent Progress in Continuous and Hybrid Reachability Analysis.* In *IEEE Conference on Computer Aided Control System Design*, pages 1582–1587, 2006.

[Ast95] Aström, K. and Wittenmark, B. *Adaptive Control.* Addison-Wesley, 1995.

[AVA19] *Transregional Collaborative Research Center "Automatic Verification and Analysis of Complex Systems (AVACS)"*, accessed 23.04.2019.
URL `http://www.avacs.org`

[Bal16] Balkan, A., Tabuada, P., Deshmukh, J. V., Jin, X. and Kapinski, J. *Underminer: A Framework for Automatically Identifying Non-converging Behaviors in Black Box System Models.* In *Proceedings of the 13th International Conference on Embedded Software*, pages 1–10, 2016.

[Bar78] Bartussek, W. and Parnas, D. L. *Using Assertions About Traces to Write Abstract Specifications for Software Modules.* Proceedings of the 2nd Conference of the European Cooperation on Informatics: Information Systems Methodology, pages 211–236, 1978.

[Bar05] Barnett, M., Rustan, K., Leino, M. and Schulte, W. *The Spec# Programming System: An Overview.* In *Construction and Analysis of Safe, Secure and Interoperable Smart Devices*, Lecture Notes in Computer Science, pages 49–69. Springer, 2005.

[Bar13] Bartoo, G. and Bogucki, T. *Essentials of Usability Engineering in Point-of-care Devices.* In *IEEE Point-of-Care Healthcare Technologies (PHT)*, pages 184–187, 2013.

[Bar18] Bartocci, E., Deshmukh, J., Donzé, A., Fainekos, G., Maler, O., Ničković, D. and Sankaranarayanan, S. *Specification-Based Monitoring of Cyber-Physical Systems: A Survey on Theory, Tools and Applications.* Springer, 2018.

[Bau87] Bauch, H., Jahn, K.-U., Oelschlägel, D., Süsse, H. and Wiebigke, V. *Intervallarithmethik: Theorie und Anwendungen.* Teubner, 1987.

[Bee72] Beeck, H. *On the Structure and Estimations of the Solution-set of a System of Linear Equations with Interval Coefficients.* Computing, volume 10, pages 231–244, 1972.

[Bem05] Bemporad, A., Garulli, A., Paoletti, S. and Vicino, A. *A Bounded-error Approach to Piecewise Affine System Identification.* IEEE Transactions on Automatic Control, volume 50, pages 1567–1580, 2005.

[Ber08] de Berg, M., Cheong, O., van Kreveld, M. and Overmars, M. *Computational Geometry: Algorithms and Applications.* Springer, 2008.

[Bha04] Bhatia, A. and Frazzoli, E. *Incremental Search Methods for Reachability Analysis of Continuous and Hybrid Systems.* In Alur, R. and Pappas, G. (Editors), *Hybrid Systems: Computation and Control*, volume 2993 of *Lecture Notes in Computer Science*, pages 142–156. Springer, 2004.

[Bla06] Blanke, M., Kinnaert, M., Lunze, J. and Staroswiecki, M. *Diagnosis and Fault-Tolerant Control.* Springer, 2006.

[Ble10] Blesa, J., Puig, V. and Saludes, J. *Robust Fault Detection using Polytope-based Set-membership Consistency Test.* In *Conference on Control and Fault-Tolerant Systems (SysTol)*, pages 726–731, 2010.

[Ble11] Blesa, J., Puig, V. and Saludes, J. *Identification for Passive Robust Fault Detection using Zonotope-based Set-membership Approaches.* International Journal of Adaptive Control and Signal Processing, volume 25, pages 788–812, 2011.

[Boe84] Boehm, B. W. *Verifying and Validating Software Requirements and Design Specifications.* IEEE Software, volume 1, pages 75–88, 1984.

[Bor09] Borchers, S., Rumschinski, P., Bosio, S., Weismantel, R. and Findeisen, R. *A Set-based Framework for Coherent Model Invalidation and Parameter Estimation of Discrete Time Nonlinear Systems.* Proceedings of the 48th IEEE Conference on Decision and Control (CDC), pages 6786–6792, 2009.

[Bra16] Bravo, J. M., Suarez, A., Vasallo, M. and Alamo, T. *Slide Window Bounded-Error Time-varying Systems Identification.* IEEE Transactions on Automatic Control, volume 61, pages 2282–2287, 2016.

[Bro05] Broy, M., Jonsson, B., Katoen, J.-P., Leucker, M. and Pretschner, A. *Model-Based Testing of Reactive Systems: Advanced Lectures.* Lecture Notes in Computer Science. Springer, 2005.

[Bro08] Bronstein, I.N., Semendjajew, K.A., Musiol G. and Mühlig, H. *Taschenbuch der Mathematik.* Verlag Harry Deutsch, 2008.

[Cac13] Caccavale, F., Marino, A., Muscio, G. and Pierri, F. *Discrete-Time Framework for Fault Diagnosis in Robotic Manipulators.* IEEE Transactions on Control Systems Technology, volume 21, pages 1858–1873, 2013.

[Cas99] Cassandras, C. G. and Lafortune, S. *Introduction to Discrete Event Systems.* The Kluwer International Series on Discrete Event Dynamic Systems, 1999.

[Cas14] Casini, M., Garulli, A. and Vicino, A. *Feasible Parameter Set Approximation for Linear Models with Bounded Uncertain Regressors.* IEEE Transactions on Automatic Control, volume 59, pages 2910–2920, 2014.

[Che14] Chen, W., Chen, W.-T., Saif, M., Li, M.-F. and Wu, H. *Simultaneous Fault Isolation and Estimation of Lithium-Ion Batteries via Synthesized Design of Luenberger and Learning Observers.* IEEE Transactions on Control Systems Technology, volume 22, pages 290–298, 2014.

[Dan11] Dang, T. *Model-Based Testing for Embedded Systems,* chapter 14. Model-Based Testing of Hybrid Systems, pages 383–424. CRC Press, 2011.

[Die17] Diehm, G. *Identifikation des menschlichen Bewegungsverhaltens auf der Basis von Primitiven.* Ph.D. thesis, Karlsruhe Institut of Technology (KIT), KIT Scientific Publishing, 2017.

[Dja17] Djaballah, A., Chapoutot, A., Kieffer, M. and Bouissou, O. *Construction of Parametric Barrier Functions for Dynamical Systems using Interval Analysis.* Automatica, volume 78, pages 287–296, 2017.

[Don10] Donzé, A. *Breach, a Toolbox for Verification and Parameter Synthesis of Hybrid Systems.* Proceedings of the 22nd International Conference on Computer Aided Verification (CAV), pages 167–170, 2010.

[Eba14] Ebadat, A., Wahlberg, B., Hjalmarsson, H., Rojas C. R., Hägg, P. and Larsson, C. A. *Applications Oriented Input Design in Time-Domain Through Cyclic Methods.* Elsevier, 2014.

[Efi13] Efimov, D., Raïssi, T., Perruquetti W. and Zolghadri, A. *Estimation and Control of Discrete-Time LPV Systems Using Interval Observers.* In *IEEE 52nd Conference on Decision and Control (CDC)*, 2013.

[ENA19] *European Initiative to Enable Validation for Highly Automated Safe and Secure Systems "Enable-S3"*, accessed 23.04.2019.
URL `http://www.enable-s3.eu`

[Eng02] Engell, S. H. (Editor). *Modelling, Analysis and Design of Hybrid Systems.* Lecture Notes in Control and Information Sciences. Springer, 2002.

[Fau03] Faulkner, L. *Beyond the Five-user Assumption: Benefits of Increased Sample Sizes in Usability Testing.* Behavior Research Methods, Instruments, & Computers, volume 35, pages 379–383, 2003.

[Fie06] Fiedler, M., Nedoma, J., Ramík, J., Rohn, J. and Zimmermann, K. *Linear Optimization Problems with Inexact Data.* Springer, 2006.

[Fol08] Foltz, C., Schneider, N., Kausch, B., Wolf, M., Schlick, C. and Luczak, H. *Usability Engineering.* Collaborative and Distributed Chemical Engineering. From Understanding to Substantial Design Process Support: Results of the IMPROVE Project, pages 527–554, 2008.

[Föl13] Föllinger, O. *Regelungstechnik: Einführung in die Methoden und ihre Anwendung.* VDE-Verlag, 2013.

[Fos15] Foster, H. D. *Trends in Functional Verification: A 2014 Industry Study.* In *52nd ACM/EDAC/IEEE Design Automation Conference (DAC)*, pages 1–6, 2015.

[Fra16] Fraden, J. *Handbook of Modern Sensors: Physics, Designs and Applications.* Springer, 2016.

[FT03] Ferrari-Trecate, G., Muselli, M., Liberati, D. and Morari, M. *A Clustering Technique for the Identification of Piecewise Affine Systems.* Automatica, volume 39, pages 205–217, 2003.

[Fut89] Futschek, G. *Programmentwicklung und Verifikation.* Springer, 1989.

[Gho17] Ghorbani, H. R. and Ahmadzadegan, M. H. *Security Challenges in Internet of Things: Survey.* In *IEEE Conference on Wireless Sensors (ICWiSe)*, pages 1–6, 2017.

[Goo77] Goodwin, G. and Payne, R. *Dynamic System Identification: Experiment Design and Data Analysis.* Academic Press, 1977.

[Hal90] Hall, A. *Seven Myths of Formal Methods.* IEEE Software, volume 7, pages 11–19, 1990.

[Har18] Harirchi, F. and Ozay, N. *Guaranteed Model-based Fault Detection in Cyber-physical Systems: A Model Invalidation Approach.* Automatica, volume 93, pages 476–488, 2018.

[Hay86] Hayes, I. *Specification directed module testing.* IEEE Transactions on Software Engineering, volume SE-12, pages 124–133, 1986.

[Hei05] Heitmeyer, C., Archer, M., Bharadwaj, R. and Jeffords, R. *Tools for Constructing Requirements Specifications: The SCR Toolset at the Age of Ten.* International Journal of Computer Systems Science and Engineering, volume 20, pages 19–35, 2005.

[Hla14] Hladík, M. *AE Solutions and AE Solvability to General Interval Linear Systems.* ArXiv e-prints, 2014.

[Hor13] Horáček, J. and Hladík, M. *Computing Enclosures of Overdetermined Interval Linear Systems.* Reliable Computing, volume 19, pages 142–155, 2013.

[IEC10] *Functional Safety of Electrical/Electronic/Programmable Electronic Safety-Related Systems.* Standard of the International Electrotechnical Commission (IEC), 2010.

[Ing09] Ingimundarson, A., Bravo, J. M., Puig, V., Alamo, T. and Guerra, P. *Robust Fault Detection using Zonotope-based Set-membership Consistency Test.* International Journal of Adaptive Control and Signal Processing, volume 23, pages 311–330, 2009.

[Ise93] Isermann, R. *Fault Diagnosis of Machines via Parameter Estimation and Knowledge Processing - Tutorial Paper.* Automatica, volume 29, pages 815–835, 1993.

[Ise06] Isermann, R. *Fault Diagnosis Systems: An Introduction from Fault Detection to Fault Tolerance.* Springer, 2006.

[Ise10] Isermann, R. and Münchhof, M. *Identification of Dynamic Systems: An Introduction with Applications.* Advanced Textbooks in Control and Signal Processing Series. Springer, 2010.

[Ise11] Isermann, R. *Fault-diagnosis Applications: Model-based Condition Monitoring: Actuators, Drives, Machinery, Plants, Sensors and Fault-tolerant Systems.* Springer, 2011.

[ISO11] *ISO 26262 - Road Vehicles - Functional Safety.* International Organization for Standardization, 2011.

[Jau01] Jaulin, L., Kieffer, M., Didrit, O. and Walter, E. *Applied Interval Analysis, with Examples in Parameter and State Estimation, Robust Control and Robotics.* Springer, 2001.

[Joh00] Johansson, K. H. *The Quadruple-tank Process: A Multivariable Laboratory Process with an Adjustable Zero.* IEEE Transactions on Control Systems Technology, volume 8, pages 456–465, 2000.

[Jon02] Jondral, F. and Wiesler, A. *Wahrscheinlichkeitsrechnung und stochastische Prozesse: Grundlagen für Ingenieure und Naturwissenschaftler.* Teubner, 2002.

[Jul05] Juloski, A., Weiland, S. and Heemels, W. *A Bayesian Approach to Identification of Hybrid Systems.* IEEE Transactions on Automatic Control, volume 50, pages 1520–1533, 2005.

[Kap16] Kapinski, J., Deshmukh, J. V., Jin, X., Ito, H. and Butts, K. *Simulation-Based Approaches for Verification of Embedded Control Systems: An Overview of Traditional and Advanced Modeling, Testing and Verification Techniques.* IEEE Control Systems, volume 36, pages 45–64, 2016.

[Kau80] Kaucher, E. *Interval Analysis in the Extended Interval Space IR.* In Alefeld, G. and Grigorieff, R. D. (Editors), *Fundamentals of Numerical Computation (Computer-Oriented Numerical Analysis)*, pages 33–49. Springer, 1980.

[Kha15] Khalil, H. K. *Nonlinear Control.* Pearson, 2015.

[Koc19] Kochdumper, N. and Althoff, M. *Sparse Polynomial Zonotopes: A Novel Set Representation for Reachability Analysis.* arXiv e-prints, volume 66, pages 132–145, 2019.

[Koe18] Koenig, A., Witzlsperger, K., Leutwiler, F. and Hohmann, S. *Overview of HAD Validation and Passive HAD as a Concept for Validating Highly Automated Cars.* at - Automatisierungstechnik, 2018.

[Kre95] Kreinovich, V. *Data Processing Beyond Traditional Statistics: Applications of Interval Computations. A Brief Introduction.* In *Proceedings of the International Workshop on Applications of Interval Computations*, 1995.

[Kre16] Krebs, S., Schnurr, C., Pfeifer, M., Weigold, J. and Hohmann, S. *Reduced-order Hybrid Interval Observer for Verified State Estimation of an Induction Machine.* Control Engineering Practice, volume 57, pages 157–168, 2016.

[Kre18] Krebs, S., Bächle, M. and Hohmann, S. *Coupled Boundary Interval Observer for LPV Systems Subject to Uncertainties in Input, Output and Parameters.* Automatica, volume 95, pages 426–432, 2018.

[Kup95] Kupriyanova, L. *Inner Estimation of the United Solution Set of Interval Linear Algebraic System.* Reliable Computing, volume 1, pages 15–31, 1995.

[Lak99] Lakeyev, A.V. *On Existence and Uniqeness of Solutions of Linear Algebraic Equations in Kaucher's Interval Arithmetic.* Reliable Computing, pages 53–65, 1999.

[Lak14] Lakeyev, A. *On Unboundedness of Generalized Solution Sets for Interval Linear Systems.* Reliable Computing, volume 19, 2014.

[Lau18] Lauer, F. *Global Optimization for Low-dimensional Switching Linear Regression and Bounded-error Estimation.* Automatica, volume 89, pages 73–82, 2018.

[Lax02] Lax, P. D. *Functional Analysis.* Wiley, 2002.

[Lju99] Ljung, L. *System Identification: Theory for the User.* Prentice Hall Information and System Sciences Series. Prentice Hall, 1999.

[Lüc80] Lücker, R. *Grundlagen digitaler Filter: Einführung in die Theorie linearer zeitdiskreter Systeme und Netzwerke.* Springer, 1980.

[Mac95] Macias, B. and Pulman, S. G. *A Method for Controlling the Production of Specifications in Natural Language.* The Computer Journal, volume 38, pages 310–318, 1995.

[Mah10] Mahmoud, M. S. *Switched Time-delay Systems: Stability and Control.* Springer, 2010.

[Mai16] Maiga, M., Ramdani, N., Trave-Massuyes, L. and Combastel, C. *A Comprehensive Method for Reachability Analysis of Uncertain Nonlinear Hybrid Systems.* IEEE Transactions on Automatic Control, volume 61, pages 2341–2356, 2016.

[Man10] Manchester, I. R. *Input Design for System Identification via Convex Relaxation.* 49th IEEE Conference on Decision and Control (CDC), pages 2041–2046, 2010.

[Mar99] Markov, S. *An Iterative Method for Algebraic Solution to Interval Equations.* Applied Numerical Mathematics, volume 30, pages 225–239, 1999.

[Mes10] Meslem, N., Ramdani, N. and Candau, Y. *Guaranteed Parameter Set Estimation for Monotone Dynamical Systems Using Hybrid Automata.* Reliable Computing, volume 14, pages 88–104, 2010.

[Mit07] Mitchell, I. M. *Hybrid Systems: Computation and Control.* In Bemporad, A., Bicchi, A. and Buttazzo, G. (Editors), *Hybrid Systems: Computation and Control: 10th International Workshop (HSCC)*, pages 428–443. Springer, 2007.

[Mün05] Münz, E. and Krebs, V. *Continuous Optimization Approaches to the Identification of Piecewise Affine Systems.* Proceedings of the 16th IFAC World Congress, volume 38, pages 349–354, 2005.

[Noc06] Nocedal, J. and Wright, S. J. *Numerical Optimization.* Springer, 2006.

[Oet64] Oettli, W. and Prager, W. *Compatibility of Approximate Solution of Linear Equations with Given Error Bounds for Coefficients and Right-hand Sides.* Numerische Mathematik, volume 6, pages 405–409, 1964.

[Ott18] Otten, S., Bach, J., Wohlfahrt, C., King, C., Lier, J., Schmid, H., Schmerler, S. and Sax, E. *Automated Assessment and Evaluation of Digital Test Drives.* In Zachäus, C., Müller, B. and Meyer, G. (Editors), *Advanced Microsystems for Automotive Applications*, pages 189–199. Springer, 2018.

[Oza12] Ozay, N., Sznaier, M., Lagoa, C. and Camps, O. *A Sparsification Approach to Set Membership Identification of Switched Affine Systems.* IEEE Transactions on Automatic Control, volume 57, pages 634 –648, 2012.

[Oza14] Ozay, N., Sznaier, M. and Lagoa, C. *Convex Certificates for Model (In)validation of Switched Affine Systems With Unknown Switches.* IEEE Transactions on Automatic Control, volume 59, pages 2921–2932, 2014.

[Pan17] Panchea, A. M., Chapoutot, A. and Filliat, D. *Extended Reliable Robust Motion Planners.* In *IEEE 56th Conference on Decision and Control (CDC)*, pages 1112–1117, 2017.

[Par72] Parnas, D. L. *A Technique for Software Module Specification with Examples.* Communications of the ACM, volume 15, pages 330–336, 1972.

[Par86] Parnas, D. L. and Clements, P. C. *A Rational Design Process: How and Why to Fake it.* IEEE Transactions on Software Engineering, volume SE-12, pages 251–257, 1986.

[Pui06] Puig, V., Ingimundarson, A., and Tornil, S. *Robust Fault Detection using Inverse Images of Interval Functions.* In *IFAC SAFEPROCESS*, 2006.

[Pui10] Puig, V. *Fault Diagnosis and Fault Tolerant Control using Set-membership Approaches: Application to Real Case Studies.* International Journal of Applied Mathematics and Computer Science, volume 20, pages 619–635, 2010.

[Que98] Quesada, J. F. *Lexical Object Theory: Specification Level.* Grammars, volume 1, pages 57–84, 1998.

[Raj13] Rajan, A. and Wahl, T. (Editors). *CESAR - Cost-efficient Methods and Processes for Safety-relevant Embedded Systems.* Springer, 2013.

[Ram09] Ramdani, N., Meslem, N. and Candau, Y. *A Hybrid Bounding Method for Computing an Over-Approximation for the Reachable Set of Uncertain Nonlinear Systems.* IEEE Transactions on Automatic Control, volume 54, pages 2352–2364, 2009.

[Ram17] Ramesh, S., Vogel-Heuser, B., Chang, W., Roy, D., Zhang, L. and Chakraborty, S. *INVITED: Specification, Verification and Design of Evolving Automotive Software.* In *54th ACM/EDAC/IEEE Design Automation Conference (DAC)*, pages 1–6, 2017.

[Rau06] Rauh, A., Hofer, E. and Auer, E. *ValEncIA-IVP: A Comparison with Order Initial Value Problem Solvers.* In *12th GRAMM - IMACS International Symposium on Scientific Computing, Computer Arithmetic and Validated Numerics*, 2006.

[Res07] Rescher, N. (Editor). *Error: on our Predicament when Things go Wrong.* University of Pittsburgh Press, 2007.

[Roe16] Roehm, H., Oehlerking, J., Woehrle, M. and Althoff, M. *Reachset Conformance Testing of Hybrid Automata.* In *Proceedings of the 19th International Conference on Hybrid Systems: Computation and Control*, pages 277–286, 2016.

[Roh12] Rohn, J. *A Handbook of Results on Interval Linear Problems.* Technical Report 1163, Institute of Computer Science, Academy of Sciences of the Czech Republic, 2012.

[Rze08] Rzeżuchowski, T. and Wąsowski, J. *Solutions of Fuzzy Equations Based on Kaucher Arithmetic and AE-solution Sets.* Fuzzy Sets and Systems, volume 159, pages 2116–2129, 2008.

[Sag68] Sage, A. P. *Optimum System Control.* Prentice-Hall, 1968.

[Sai14] Sainz, M., Armengol, J., Calm, R., Herrero, P., Jorba, L. and Vehi, J. *Modal Interval Analysis - New Tools for Numerical Information.* Springer, 2014.

[San17] d. Sandretto, J. A., Chapoutot, A. and Mullier, O. *Formal Verification of Robotic Behaviors in Presence of Bounded Uncertainties.* In *First IEEE International Conference on Robotic Computing (IRC)*, pages 81–88, 2017.

[Sax08] Sax, E. H. (Editor). *Automatisiertes Testen Eingebetteter Systeme in der Automobilindustrie.* Hanser, 2008.

[Sch03] Schröder, J. *Modelling, State Observation and Diagnosis of Quantised Systems.* Lecture Notes in Control and Information Sciences. Springer, 2003.

[Sch09] Schlage, T., Schwaiger, M., Krebs, V. and Lunze, J. *Comparison of Two Model-based Methods of Remote Diagnosis of Technological Systems.* at - Automatisierungstechnik, volume 57, 2009.

[Sch15a] Schätz, B., Voss, S. and Zverlov, S. *Automating Design-space Exploration: Optimal Deployment of Automotive SW-components in an ISO26262 Context.* In *52nd ACM/EDAC/IEEE Design Automation Conference (DAC)*, pages 1–6, 2015.

[Sch15b] Schupp, S., Ábrahám, E., Chen, X., Ben Makhlouf, I., Frehse, G., Sankaranarayanan, S. and Kowalewski, S. *Current Challenges in the Verification of Hybrid Systems.* Cyber Physical Systems. Design, Modeling, and Evaluation: 5th International Workshop, pages 8–24, 2015.

[Sha96] Shary, S. *Algebraic Approach to the Interval Linear Static Identification, Tolerance and Control Problems or one more Application of Kaucher Arithmetic.* Reliable Computing, volume 2, pages 3–33, 1996.

[Sha02] Shary, S. *A New Technique in Systems Analysis under Interval Uncertainty and Ambiguity.* Reliable Computing, volume 8, pages 321–418, 2002.

[Sha10] Shary, S. *A New Method for Inner Estimation of Solution Sets to Interval Linear Systems.* In Rauh, A. and Auer, E. (Editors), *Modeling, Design and Simulation of Systems with Uncertainties*, pages 21–42. Springer, 2010.

[Sha14] Shary, S. *On Full-Rank Interval Matrices.* Numerical Analysis and Applications, volume 7, pages 241–254, 2014.

[Spi89] Spivey, J. M. *An Introduction to Z and Formal Specifications.* Software Engineering Journal, volume 4, pages 40–50, 1989.

[TF91] Thevenod-Fosse, P., Waeselynck, H. and Crouzet, Y. *An Experimental Study on Software Structural Testing: Deterministic Versus Random Input Generation.* In *Twenty-First International Symposium on Fault-Tolerant Computing*, pages 410–417, 1991.

[Tip00] Tipler, P. A. and Walker, J. S. *Physik.* Spektrum Akademischer Verlag, 2000.

[Uga03] Ugarte, I. and Sanchez, P. *Functional Vector Generation for Assertion-based Verification at Behavioral Level using Interval Analysis.* In *Eighth IEEE International High-Level Design Validation and Test Workshop*, pages 102–107, 2003.

[Utt06] Utting, M., Pretschner, A. and Legeard, B. *A Taxonomy of Model-Based Testing.* Working Paper Series, 2006.

[Ven03a] Venkatasubramanian, V., Rengaswamy, R., Yin, K. and Kavuri, S. N. *A Review of Process Fault Detection and Diagnosis: Part I: Quantitative Model-based Methods.* Computers & Chemical Engineering, volume 27, pages 293–311, 2003.

[Ven03b] Venkatasubramanian, V., Rengaswamy, R. and Kavuri, S. N. *A Review of Process Fault Detection and Diagnosis: Part II: Qualitative Models and Search Strategies.* Computers & Chemical Engineering, volume 27, pages 313–326, 2003.

[Ven03c] Venkatasubramanian, V., Rengaswamy, R., Kavuri, S. N. and Yin, K. *A Review of Process Fault Detection and Diagnosis: Part III: Process History Based Methods.* Computers & Chemical Engineering, volume 27, pages 327–346, 2003.

[Ven15] Vento, J., Blesa, J., Puig, V. and Sarrate, R. *Set-membership Parity Space Hybrid System Diagnosis.* International Journal of Systems Science, volume 46, pages 790–807, 2015.

[Vid08] Vidal, R. *Recursive Identification of Switched ARX Systems.* Automatica, volume 44, pages 2274–2287, 2008.

[Wac17] Wachenfeld, W. H. K. *How Stochastic can Help to Introduce Automated Driving.* Ph.D. thesis, Technische Universität Darmstadt, 2017.

[Wan17] Wang, H., Kolmanovsky, I. and Sun, J. *Zonotope-based Set-membership Parameter Identification of Linear Systems with Additive and Multiplicative Uncertainties: A new Algorithm.* In *American Control Conference (ACC)*, pages 1481–1486, 2017.

[Wan18] Wang, Y., Puig, V. and Cembrano, G. *Set-membership Approach and Kalman Observer Based on Zonotopes for Discrete-time Descriptor Systems.* Automatica, volume 93, pages 435–443, 2018.

[Web09] Weber, J. *Automotive Development Processes: Processes for Successful Customer Oriented Vehicle Development.* Springer, 2009.

[Wil64] Wilde, D. J. *Optimum Seeking Methods.* Prentice-Hall, 1964.

[Wil17] Williams, R. S. *What's Next? [The end of Moore's law].* Computing in Science Engineering, volume 19, pages 7–13, 2017.

[Wol10] Wolff, F. *Konsistenzbasierte Fehlerdiagnose nichtlinearer Systeme mittels Zustandsmengenbeobachtung.* Ph.D. thesis, Karlsruhe Institut of Technology (KIT), KIT Scientific Publishing, 2010.

[Zai14] Zaiser, S., Buchholz, M. and Dietmayer, K. *Interval System Identification for MIMO ARX Models of Minimal Order.* 53rd IEEE Conference on Decision and Control, pages 1774–1779, 2014.

[Zan12] Zander, Justyna, Schieferdecker, Ina and Mosterman, Pieter J. (Editors). *Model-Based Testing for Embedded Systems.* Taylor & Francis, 2012.

[ZN09] Zander-Nowicka, J. *Model-based Testing of Real-time Embedded Systems in the Automotive Domain.* Ph.D. thesis, Technische Universität Berlin, 2009.

[Zol14] Zolghadri, A., Henry, D., Cieslak, J., Efimov, D. and Goupil, P. *Fault Diagnosis and Fault-Tolerant Control and Guidance for Aerospace Vehicles.* Springer, 2014.

Own Publications and Conference Contributions

[Die13a] Diehm, G., Maier, S., Flad, M. and Hohmann, S. *An Identification Method for Individual Driver Steering Behaviour Modelled by Switched Affine Systems.* In *52nd IEEE Conference on Decision and Control (CDC)*, pages 3547–3553, 2013.

[Die13b] Diehm, G., Maier, S., Flad, M. and Hohmann, S. *Online Identification of Individual Driver Steering Behaviour and Experimental Results.* In *IEEE International Conference on Systems, Man, and Cybernetics (SMC)*, pages 221–227, 2013.

[Fla14a] Flad, M., Otten, J., Schwab, S. and Hohmann, S. *Necessary and Sufficient Conditions for the Design of Cooperative Shared Control.* In *IEEE International Conference on Systems, Man, and Cybernetics (SMC)*, pages 1253–1259, 2014.

[Fla14b] Flad, M., Otten, J., Schwab, S. and Hohmann, S. *Steering Driver Assistance Aystem: A Systematic Cooperative Shared Control Design Approach.* In *IEEE International Conference on Systems, Man, and Cybernetics (SMC)*, pages 3585–3592, 2014.

[Lem18] Lemmer, M., Köpf, F., Schwab, S., Flad, M. and Hohmann, S. *Modeling of Human-Centered Cooperative Control by Means of Tracking in Discrete Time Linear Quadratic Differential Games.* In *Proceedings of Artificial Intelligence and Knowledge Engineering*, 2018.

[Sch16] Schwab, S. and Hohmann, S. *Verification of Battery Management Systems using Hybrid Identification.* In *SINO-EU Doctoral School for Sustainability Engineering*, 2016.

[Sch17a] Schwab, S., Holzmüller, B. and Hohmann, S. *Automated Verification of Switched Systems Using Hybrid Identification*, pages 87–100. Springer, 2017.

[Sch17b] Schwab, S., Stark, O. and Hohmann, S. *Examples on Verified Diagnosis of Safety Critical Dynamic Systems Based on Kaucher Interval Arithmetik.* In *10th Summer Workshop on Interval Methods*, 2017.

[Sch17c] Schwab, S., Stark, O. and Hohmann, S. *Verified Diagnosis of Safety Critical Dynamic Systems Based on Kaucher Interval Arithmetic.* Proceedings of the 20th IFAC World Congress, 2017.

[Sch18a] Schwab, S. and Hohmann, S. *Automatisierte Verifikation hybrider Systeme am Beispiel eines Batteriemanagementsystems.* 15. Fachtagung EKA-Entwurf komplexer Automatisierungssysteme, 2018.

[Sch18b] Schwab, S., Puig, V. and Hohmann, S. *A Robust Fault Detection Method using a Zonotopic Kaucher Set-membership Approach.* 10th IFAC Symposium on Fault Detection, Supervision and Safety for Technical Processes (SAFEPROCESS), 2018.

[Sch18c] Schwab, S., Puig, V. and Hohmann, S. *A Robust Fault Detection Method using a Zonotopic Kaucher Set-membership Approach - Application to a Real Single-Tank Process.* In *11th Summer Workshop on Interval Methods*, 2018.

[Sch19] Schwab, S. and Hohmann, S. *Verification of Hybrid Systems Using Kaucher Arithmetic.* at - automatisierungstechnik, volume 67, pages 316–325, 2019.

[Var19] Varga, B., Meier, S., Schwab, S. and Hohmann, S. *Model Predictive Control and Trajectory Optimization of Large Vehicle-manipulators.* IEEE International Conference on Mechatronics, 2019.

Supervised Theses

[Adi16] Adiraju, S. *Optimization Based Verification for Battery Management Systems.* Master thesis, Faculty of Electrical Engineering, Karlsruhe Institute of Technologie (KIT), 2016.

[Glü17] Glück, L. *Modellierung und Analyse eines hybriden Bremssystems.* Bachelor thesis, Faculty of Electrical Engineering, Karlsruhe Institute of Technologie (KIT), 2017.

[Hen15] Henn, Y. *Kombinierte Verifikation von dynamischen Subsystemen und ihrem überlagerten Schaltautomaten.* Master thesis, Faculty of Electrical Engineering, Karlsruhe Institute of Technologie (KIT), 2015.

[Kah15] Kahraman, B. *Definition und Implementation eines trajektorienbasierten Spezifikationsverfahrens.* Diploma thesis, Faculty of Electrical Engineering, Karlsruhe Institute of Technologie (KIT), 2015.

[Lem15] Lemmer, M. *Aufbau einer BMS Testumgebung zur garantierten Verifikation durch hybride Identifikation.* Bachelor thesis, Faculty of Electrical Engineering, Karlsruhe Institute of Technologie (KIT), 2015.

[Rie17] Rieser, L. *Inputdesign für ein neuartiges Verfahren zur Verifikation von Cyber-Physical Systems.* Master thesis, Faculty of Electrical Engineering, Karlsruhe Institute of Technologie (KIT), 2017.

[Sta16] Stark, O. *Optimierungsbasierte Verfahren zur intervallhaften Identifikation.* Master thesis, Faculty of Electrical Engineering, Karlsruhe Institute of Technologie (KIT), 2016.

[Thi16] Thimm, M. *Identifikation hybrider Systeme mit einem intervallhaften Modulationsfunktionsverfahren.* Master thesis, Faculty of Electrical Engineering, Karlsruhe Institute of Technologie (KIT), 2016.

[ZO14] Zotes Orcajo, A. G. *Implementierung eines hybriden Batteriemanagementmodells.* Bachelor thesis, Faculty of Electrical Engineering, Karlsruhe Institute of Technologie (KIT), 2014.

Karlsruher Beiträge zur Regelungs- und Steuerungstechnik (ISSN 2511-6312)
Institut für Regelungs- und Steuerungssysteme

Band 01 Diehm, Gunter
Identifikation des menschlichen Bewegungsverhaltens auf der Basis von Primitiven.
ISBN 978-3-7315-0608-9

Band 02 Flad, Michael
Kooperative Regelungskonzepte auf Basis der Spieltheorie und deren Anwendung auf Fahrerassistenzsysteme.
ISBN 978-3-7315-0610-2

Band 03 Eckert, Marius
Modellbasierte Identifikation fraktionaler Systeme und ihre Anwendung auf die Lithium-Ionen-Zelle.
ISBN 978-3-7315-0690-4

Band 04 Krebs, Stefan
Intervallbeobachter für lineare parametervariante Systeme und deren Anwendung auf die Asynchronmaschine.
ISBN 978-3-7315-0857-1

Band 05 Kaspar, Stephan
Fahrdynamikuntersuchungen eines Elektrofahrzeugs mit Einzelrad-Hinterradantrieb.
ISBN 978-3-7315-0916-5

Band 06 Sauter, Patrick S.
Modellierung und zentrale prädiktive Regelung von multimodalen Energieverteilnetzen.
ISBN 978-3-7315-0963-9

Band 07 Kupper, Martin
Verteilte Zustandsschätzung fraktionaler Systeme und ihre Anwendung auf Lithium-Ionen-Batteriesysteme.
ISBN 978-3-7315-0971-4

Band 08 Merkert, Lennart
Optimal Scheduling of Combined Heat and Power Generation Considering Heating Grid Dynamics.
ISBN 978-3-7315-1056-7

Die Bände sind unter www.ksp.kit.edu als PDF frei verfügbar oder als Druckausgabe bestellbar.

Band 09 Ludwig, Julian
Automatisierte kooperative Transition einer Regelungsaufgabe zwischen Mensch und Maschine am Beispiel des hochautomatisierten Fahrens.
ISBN 978-3-7315-1069-7

Band 10 Inga Charaja, Juan Jairo
Inverse Dynamic Game Methods for Identification of Cooperative System Behavior.
ISBN 978-3-7315-1080-2

Band 11 Schnurr, Christoph Xaver
Ein Verfahren zur lexikographischen modellprädiktiven Regelung mit der Anwendung auf eine permanenterregte Synchronmaschine.
ISBN 978-3-7315-1095-6

Band 12 Schwab, Stefan
Guaranteed Verification of Dynamic Systems.
ISBN 978-3-7315-0965-3